Winning through Innovation

Winning through Innovation

Lessons from the Front
Lines of Business

By
Stephen W. Hinch

**ATENERA
PRESS**

Atenera Press · Santa Rosa, California

ATENERA PRESS

Atenera Press
Santa Rosa, California
publisher@atenera.com
www.atenera.com

ISBN: 978-0-9661999-1-8 (hardcover)
ISBN: 978-0-9661999-4-9 (paperback)
ISBN: 978-0-9661999-5-6 (ebook)
Library of Congress Control Number: 2025901298

Table of Contents

Foreword...ix
Introduction xi

Part I: What Is Innovation?...1
Chapter 1:   The Innovative Organization 3
Chapter 2:   Innovation as a Competitive Weapon 21
Chapter 3:   Fortune 100 Case Study: Product Innovation
             at Hewlett-Packard Company................................39
Chapter 4:   SMB Case Study: Process Innovation at
             Carbon Systems...65

Part II: Empowering Incremental and Disruptive Innovation 79
Chapter 5:   Creating an Environment for Incremental Innovation 81
Chapter 6:   Historical Case Study: Disruptive Innovation in the
             Railroad Industry 103
Chapter 7:   Creating an Environment for Disruptive Innovation 123
Chapter 8:   Case Study: Special Case of Disruptive Innovation...........149

Part III: Essential Dynamics of Innovation 157

Chapter 9: Market-Driven Innovation 159

Chapter 10: Innovation as a Function of Market Phase 175

Chapter 11: Driving Innovation by Influencing the Industry 189

Chapter 12: Innovation in the Digital Age 199

Acknowledgments 211

About the Author 215

References 217

Index 221

To managers everywhere who must balance the needs of investors, executives, customers, and employees without mortgaging the future in the quest for short-term results.

Foreword

It is with great pleasure that I introduce *Winning through Innovation* by Steve Hinch, a distinguished technology executive, IT managed service provider, and former TeamLogic IT franchisee. Steve's extensive experience and deep understanding of entrepreneurship and innovation make this book an invaluable resource for anyone looking to navigate the complexities of managing a tech venture.

In the ever-evolving landscape of information technology, innovation is not just a buzzword; it is a necessity. I lead a company, TeamLogic IT, that provides advanced technology solutions for companies of all sizes—services such as managed IT support, business continuity, cloud services, and technology consulting. As a company on the front lines of technology, we know how essential innovation is to our success. It keeps us ahead of cybersecurity threats to our clients and allows us to take advantage of carefully tested technology advancements such as intelligent automation and artificial intelligence. In the world of IT, one can never remain static. Steve's experience, insights, and judgment have provided me with valuable guidance helping TeamLogic IT develop and execute strategic initiatives.

Steve's journey in the technology industry is nothing short of inspiring. Steve built his first career responsible for R&D at industry pioneer

Hewlett-Packard. As a former TeamLogic IT franchisee, he has firsthand experience in managing and growing IT services in a competitive market. His transition from a franchisee to a consultant has allowed him to gain a broader perspective on the industry, making his insights even more valuable. Steve's ability to adapt and innovate has been a key factor in his success, and this book is a testament to his expertise.

Winning through Innovation is not just a guide; it is a road map for success in industry. Steve's approach to innovation is both practical and visionary. He emphasizes the importance of staying ahead of the curve and continuously seeking new ways to improve and evolve. His strategies are grounded in real-world experience, making them both relatable and actionable.

This book reflects Steve's commitment to helping others succeed. His passion for innovation shines through in every page of this book. He understands that innovation is not just about adopting the latest technologies; it is about fostering a culture of continuous improvement and creativity. His insights on how to cultivate an innovative mindset are both inspiring and practical, providing readers with the tools they need to drive change within their own organizations.

As you read *Winning through Innovation*, you will be struck by Steve's ability to distill complex concepts into clear, actionable steps. His writing is both engaging and informative, making it easy to absorb and apply his insights. This book is not just a theoretical exploration of innovation; it is a practical guide that you can use to achieve tangible results. *Winning through Innovation* is a must-read for anyone looking to thrive in the technology industry or for anyone starting a new business.

—Dan Shapero
President and COO, TeamLogic Inc.

Introduction

Welcome, readers, to this book about innovation in the business world. Innovation has been a hot topic for decades, and you might ask how this book is different from all the others already out there. To answer that question, let's start with a story. For those of you already in management, it may sound familiar. For those of you who are not yet there, its message may help make you a better manager when the opportunity arises.

When I took on my first role as a manager at Hewlett-Packard Company after several years as an engineer, I quickly discovered there was a gap between the management theories I had learned in school and the management challenges I faced on the front lines of business. The theories were sound, but they couldn't teach me everything I needed to know. My first months on the job were a struggle.

My own manager coached me through these struggles by stressing the importance of supplementing theory with guidance from more experienced managers who had already faced challenges similar to mine. He impressed on me the importance of blending management theory with practical insight.

As I progressed up the ranks of management, I found this same concept applied to innovation. Many excellent books had been written about innovation. Invariably, they dealt with it from the high-level perspective of an academic or a consultant. What I found missing were detailed stories from real managers who shared what they did to drive innovation and deal with the challenges they faced along the way.

I wrote *Winning through Innovation* to offer this new level of insight. I have led many innovation programs, including incremental and disruptive innovations for both products and processes. Over the course of this book, I will draw on many of these experiences as examples. Not all of them are positive. Sometimes the best learning experiences come from failure.

WHAT MAKES THIS BOOK DIFFERENT

Amid the growing field of innovation literature, what sets this book apart? Foremost, it is a practical guide, not an academic treatise. Many of the case studies I share are ones my colleagues and I have lived and breathed for years at a time. As such, I can go into more depth than would be possible from a consultant or academic professor. Consultants may spend a few weeks or months coaching managers on how to innovate, but they do not live or die by the outcome the same way the manager does. To quote the old saw that in a ham-and-eggs breakfast, the chicken is involved while the pig is committed—the consultant is involved while the manager is committed.

How will you see this play out in the chapters that follow? I have built this book around extended, unpolished case studies that I use to illustrate the key concepts of innovation. In these case studies, you will learn not only the actions that managers took and the results they achieved but also the thought processes they went through to get there. Unlike the simple, cleaned-up case studies you typically read, these are messy and cluttered—just like real life. No single person can take credit for the results, be they good or bad. Not every decision turned out well, but there are lessons to be learned from all of them. To quote another old saw attributed to everyone from Mark Twain to Will Rogers, "Good judgment comes from experience. Experience comes from bad judgment." My colleagues and I can safely claim to have a lot of experience.

I draw heavily from my own multidecade career at Hewlett-Packard, Agilent Technologies, and TeamLogic IT. As a midlevel manager for many of those years, I understand the challenges faced by those managers. They must negotiate a balance between the needs of their businesses and the demands of senior management, which do not always align.

So why this book at this moment in time? As we all know, times are changing in the business world at what can feel like an unprecedented clip. The Digital Age has brought innovations such as generative artificial intelligence into the imagination of business leaders everywhere. The natural question is "How is innovation different today than it was ten years ago?"

In the chapters that follow, I will show that the basic concepts have not changed—a fact that managers at all levels who are reading this book will hopefully consider good news! What the Digital Age brings is a set of tools that, if used properly, deliver new value to the innovation process. As long as you understand the core tenets of innovation, you will be able to use these tools to be even more successful.

My approach is guided by what I call *high-contrast management*, a philosophy of simplification meant to deliver a clear, easily understandable vision to the relevant parties—the actual "users" of innovation. As such, I focus on the concepts necessary to make innovation a competitive weapon rather than focusing on the details of specific tools. The most popular software tools to help with innovation may change every year, but the principles behind how to innovate do not. If you understand these, you will be prepared for the future regardless of what management tools or software programs are in vogue at the moment.

HOW THIS BOOK CAN WORK FOR YOU

I wrote *Winning through Innovation* to be useful to a range of readers across a variety of industries. Target readers include:

- Midlevel managers in the high technology industry
- Senior leaders in the high technology industry

- Senior and midlevel managers in other industries who want to drive innovation
- Leaders in small and medium-size businesses (SMBs) who want to drive innovation
- First-level managers and individual contributors across all industries who want to learn how they can contribute to innovation

There is something here for all these readers. Midlevel managers will learn how others in similar roles have navigated through the competing demands of executives, customers, employees, and investors to drive innovation success. Senior leaders will learn how others have created working environments that allow those midlevel managers to be successful. First-level managers and individual contributors will learn how executives think about innovation and the kinds of innovative ideas they value. Leaders in small businesses will learn that innovation is not limited to large corporations and venture-capital-funded start-ups.

Reading this book can only be your first step in creating an innovative environment. Innovation flourishes when everyone on the team understands what it takes to be successful. To help with that goal, I end each chapter with a set of questions that cover topics in that chapter. I encourage team members to first read each chapter on their own, then gather in a group to discuss these questions. There are no absolute right or wrong answers, and the best answer may be different depending on the company.

HIGH-CONTRAST MANAGEMENT

Throughout my career, I have always considered it important to simplify complex issues so they are easier to understand and act on. I have seen too many managers in high tech agonize endlessly over a decision. It comes from their inherently analytic nature. Engineers are taught from the first day of their first engineering class to analyze a problem from every angle. Do not be quick to jump to a conclusion; take the time to think it through and get the right answer.

For an engineer, that is sound logic. For a manager, it is a recipe for paralysis. There are few "right" answers in management. Every decision will have trade-offs. Overanalyzing it will not make those trade-offs go away. Rather than searching endlessly for a decision without trade-offs, acknowledge the trade-offs, make the decision, and move on. This ability to frame decisions in high contrast and act quickly becomes ever more important the higher you move in management. General George Patton said it best: "A good decision today is better than the perfect decision next week."

This book reflects that philosophy of simplification. Some readers may feel at times it is too simplistic. For example, in my discussion in chapter 1 of the various types of innovation, I limit it to four categories. This is far fewer than other authors, who may classify innovation into a dozen or more categories. You may ask that if respected experts claim there are as many as fourteen types of innovation, why should Steve Hinch feel he can reduce it down to four?

The answer is that as a manager, my job is to make fast, accurate decisions with less data than I wish I had. The way to do this is to simplify things so I can more easily make sense of them. Only then will I have the insight necessary to make a good decision. My approach is to start by understanding the big picture. I add complexity later, when it is more easily digested.

This approach not only helps my own thought process but is also essential when selling a vision across the organization. The typical corporate vice president or CEO did not rise to their position because of the depth of their technical curiosity. If that were what drove them, they would still be engineers. They got there because of their business acumen. That means they will be comfortable making important decisions in real time based on limited data. Overly complicated, in-depth analyses do not impress them as much as clear, concise reasoning. Try explaining your strategy in the context of fourteen types of innovation, and their eyes will glaze over within the first five minutes. Position it in a simpler four-quadrant framework, and it is a much easier message to absorb. It is no accident that the well-known Boston Consulting Group's model for portfolio management also uses only four quadrants.

I have watched otherwise brilliant people fail to understand this point. They create lengthy presentations filled with intricate details from the latest thinking out of academic circles, confident they will impress company executives. When they don't, they leave frustrated, claiming that management "doesn't get it." But the problem does not lie with the executives. If the presenter had only understood the needs of the audience and crafted a message tailored to that audience, the outcome would have been different. In business, simplified, easy-to-understand messages work best.

That is not to say the more complex analysis is unimportant. Do not hesitate to seek advice from those consultants and academic experts. They can bring valuable insight that will undoubtedly help you craft a better solution. Use their guidance as you think through your strategy and develop your plans. Then simplify your message so the many others you will rely on for execution—both above you and below you in the organization—can more easily absorb it. Remember, your audience will not have the benefit of having analyzed the subject in depth the same way you have. Concepts that seem simple after hours of thought will not feel that way to someone learning about them for the first time.

Long after the consultants have gone home and the management books are back on the shelf, managers are the ones who are left accountable for the performance of their businesses. They can read all the books and talk to all the consultants they want. But unlike law, where following the advice of your attorney is a legitimate defense, explaining away poor business results by claiming you "followed the advice of your consultant" won't cut it. Use that excuse and you'd better have your résumé up to date. If the insight conveyed in this book allows you to avoid such a situation, it will have accomplished its goal.

Part I
What Is Innovation?

Chapter 1

The Innovative Organization

Every manager in the business world has heard the dire warnings. Innovate or die. Obsolete yourself before your competition does it for you. Only the paranoid survive. Fearful of being left behind, every company strives to out-innovate its competitors to the extent that you'd think society would be awash in unbridled growth.

And yet, most of that innovation fails to fulfill its promise. For every Apple iPhone or Tesla EV, a thousand other innovations end up on the scrap heap of failed ideas. How can this be? Academic texts on innovation and growth have been around for decades, and the theories they promote are sound.

The challenge is that management and innovation happen in real life. If a manager could only devote all her attention to nurturing innovation, life would be so much easier. But stark reality intrudes. Investors demand consistent quarterly performance. Executives have little patience for a manager who misses his numbers to fund an innovative new idea. The finance team seems more interested in reducing research-and-development (R&D) expenses and increasing gross margins than in helping build a business case for that new idea. Customers complain about defects in current products that are making their lives miserable and must be fixed. The real-life manager must deal with all this while simultaneously trying to position the product line for long-term growth.

INNOVATION DEFINED

Today, the importance of innovation to a company's success is broadly accepted. But what exactly does it mean to be innovative? A review of the literature provides many definitions, generally along the lines of "inventing something new." In the business world, that is not sufficient. Innovation requires more than just invention. Here is a better definition:

> *Innovation is the ability to see opportunity in places others do not and turn that vision into profitable reality.*

Notice the two parts to this definition: the ability to see opportunity—the innovative idea—and the ability to turn that idea into profitable reality. You need to be good at both. In business, good ideas are worth nothing if they are not successfully commercialized. That is why so much of this book is devoted to execution—the "D" part of R&D.

The tendency in many companies is to think about these two parts separately. For example, the senior leadership team of a business may get together for regular "innovation" discussions. From these come such things as market segmentation reports and product portfolio plans. But how often do these discussions include an assessment of the organization's ability to deliver? If every new product or service introduced in the last three years took 30 percent longer to develop than originally planned, has the leadership team added an objective for improving execution?

It should. What good is it to come up with innovative ideas if you cannot convert them into a sustainable benefit? Conversely, what good is it to introduce new products or services on time and on budget if you have not chosen the right ones to introduce?

THE FIVE TENETS OF INNOVATION

One of my primary aims in this book is to reassure you that innovation is not a mysterious process. Let's start with what I consider the five most important tenets of innovation. They are especially applicable to companies in the world of high tech, but the concepts apply across a much broader range of businesses:

1. Any company that wants to grow needs to be good at innovation.
2. There are various types of innovation, and the best companies excel at all of them.
3. Innovation requires commitment from all levels of the organization, starting at the top.
4. The R&D function must be an innovation engine.
5. Innovation is a team game; R&D cannot do it all on its own.

At this point, some readers may say, "That would be great if I worked in high tech, but I work in an attorney's office. We don't have an R&D department. Does that mean we can't innovate?" Not at all. The attorney who says "I want to figure out how to use artificial intelligence to improve the quality of our contracts" is innovating just as much as the R&D engineer at Microsoft. They need to be given similar support for pursuing their ideas. Do not get hung up on organization structures being impediments to innovation.

Let's explore each of these tenets in more detail.

1. Any company that wants to grow needs to be good at innovation.

Two modern examples illustrate the difference between a market driven by innovation and one that is not. To remain successful, companies must recognize and be responsive to an ever-changing business landscape.

First, think about the market for gasoline.[1] When you fill up your car's tank, what brand are you likely to choose? The answer most people give is "Whichever one is cheapest!" We all know people who drive all the way across town to save a penny or two a gallon, oblivious to the fact the extra gas they consume getting there more than offsets any savings. Chevron, Shell, BP, and ExxonMobil all have a tough challenge. In the mind of the consumer, the main differentiation among brands is price. Their primary avenue for growth is to raise prices on the assumption that their competition will do the same. At the moment there is enough business for all of them, but that will change as the automobile industry moves to electric. As that happens, companies that know how to innovate will out-compete those that don't.

Now think about smartphones. What brands come to the top of your mind? Apple's iPhone? Google's Android? What about Motorola, Nokia, or Blackberry? How many people even remember that Blackberry was once the most popular smartphone? Apple has established such a strong perception as the innovation leader that the iPhone can command a premium price even in the face of widespread competition.

Why is it that some companies are successful innovators while others are not? The first thing to understand is that annual surveys of "The World's Most Innovative Companies" do not tell the whole story. Such surveys invariably confuse innovative ideas with innovative companies. Any company can come up with a single innovative idea. What distinguishes a truly innovative company is that it has installed processes that allow it to repeat that success more than once. Sorare may well deserve to be on the same list as Apple someday, but I would want to see a more sustained track record first. Any company on a "Most Innovative Companies" list that has not demonstrated an ongoing, multiyear track record of innovation should not be there. Put it on a "Ones to Watch" list instead.

The secret to successful innovation is that you do not just turn people loose and wait to reap the rewards. You need to approach it methodically.

---

[1] "Petrol" for those of you in the UK and Commonwealth countries.

When I first became an R&D manager, people told me, "You can't schedule invention. R&D engineers will take as long as necessary to get the job done, and you can't predict in advance how long that will be." As Einstein allegedly said, "If we knew what we were doing, it wouldn't be research, would it?"

I soon discovered this mantra was little more than a way to cover up poor planning. Most R&D activities—perhaps as much as 80 percent—are predictable tasks that can be readily scheduled. The other 20 percent—the "R" part of R&D—typically comes at the very beginning of the project before you have committed to a firm schedule. This is the time when you are inventing the few differentiating technologies that will be key to the product. This period of invention may not lend itself to precise scheduling, but by confining it to the earliest phase of a project, you can keep the investment under control.

This does not mean that all uncertainty is limited to the earliest phase of a project. Unpredictable events will certainly occur later on, such as when a circuit design fails to meet its performance requirements or a bug-filled software program crashes repeatedly. But these uncertainties can be accommodated in the schedule through time-proven statistical methods.

What does this have to do with innovation? Remember that innovation is not just coming up with novel ideas. It also means turning those ideas into reality. To be good at innovation, a company must be good at execution. The best companies know how to embed these concepts into company culture so they are not dependent on luck or the skills of a few "supersmart" individuals.

2. There are various types of innovation, and the best companies excel at all of them.

Many books divide innovation into numerous categories.[2] That makes sense from an academic perspective, but as an industry executive, I have found it useful to think in much simpler terms.

---

[2] Some authors talk about as many as ten types of innovation. Geoffrey Moore, in his book, *Dealing with Darwin*, lists no fewer that fourteen.

Imagine innovation being structured along two orthogonal axes. On the horizontal axis, define it as either product innovation or process innovation. On the vertical axis, define it as either incremental innovation or disruptive innovation. The result, as shown in Figure 1.1, is a square with four quadrants.

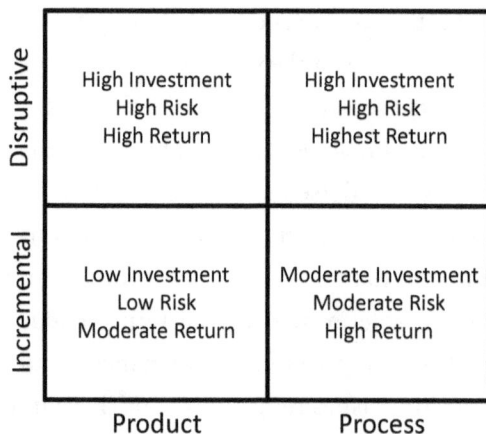

	Product	Process
Disruptive	High Investment High Risk High Return	High Investment High Risk Highest Return
Incremental	Low Investment Low Risk Moderate Return	Moderate Investment Moderate Risk High Return

Figure 1.1: The four main categories of innovation

Each quadrant presents a general sense of the relative level of investment, risk, and return for that type of innovation. For example, a disruptive process innovation can return large rewards, but to achieve it you must make a large investment and accept a high level of risk. This suggests you should only make such an investment when the potential return is also high. There may be considerable risk in achieving that return, but the potential must be there.

My first opportunity to lead a disruptive process innovation came when I was hired by HP's Corporate Manufacturing Engineering department to lead the company's migration from an older, heavily manual process for assembling printed circuit boards to one that was fully automated. It took almost four years and required us to invest many millions of dollars in engineering and capital equipment, but when the migration was complete, HP's new process was so reliable and efficient that it was unparalleled in the industry.

An incremental product innovation is unlikely to generate large returns, but it is much less risky. It is usually a less expensive undertaking, so it often makes sense to do even if the returns are moderate.

Automobile manufacturing is an example of an industry that regularly deploys incremental product innovations. Manufacturers do not hesitate to redesign vehicle components to save as little as a dollar or two per car because over the hundreds of thousands of cars they manufacture every year, the savings can be substantial. Of course, these are generalizations; "Your mileage may vary."

Note that I do not classify any of the quadrants as "low return." All types of innovation can be worthwhile under the right circumstances. Calling any of the quadrants "low return" would incorrectly imply that this kind of innovation is not a good way to invest. Not every quadrant makes sense at any given time, but none of the quadrants are inherently poor choices. The only low-return choice is no innovation at all.

Now let's explore the horizontal axis in more detail. For this simplified view of innovation, think of "product" in broad terms that also include "services." Companies like Sony or Procter & Gamble sell physical products you can hold and touch, whether they be high-definition TVs or tubes of toothpaste. Companies like United Airlines or Dropbox sell services, not physical products. The service United Airlines sells you is transportation from one city to another. The service Dropbox sells you is the right to use their cloud-based services for your IT needs. For our purposes, the single term "product" will include both services and physical products.

Similarly, think of "process" in broad terms. A process includes not only the methodology used to produce a product but also the channel used to deliver that product to the customer. Amazon's sales channel innovation, for example, was a new service made possible by a process innovation that revolutionized retailing. It made online purchasing safe, convenient, and less expensive than a traditional brick-and-mortar store.

Google started with a business model of giving away all their services for free, funded entirely by advertising. Although not exactly original (network television has done it that way for decades), it has certainly been one of the more successful process models of the internet era.

Product innovation tends to grab the headlines because it results in something readily apparent to customers—BMW's latest sports car will certainly create a buzz both in the press and with car buyers. Process innovation, though, can be even more important because it benefits every product that goes through that process. Improving the manufacturing process for BMW automobiles, for example, will benefit every vehicle that goes through it, not just the innovative new sports car hyped in the press. This is one reason it is dangerous to gauge a company's ability to innovate by the number of patents it produces; few companies patent process improvements nearly as often as product improvements.[3]

Now let's turn our attention to the chart's vertical axis. Clayton Christensen used the terms *sustaining innovation* and *disruptive innovation* in his classic book *The Innovator's Dilemma*. In it he defined sustaining innovations as those that improve existing product performance. Disruptive innovations fundamentally change markets. Other authors use the terms *incremental innovation* and *radical innovation*. Incremental innovation builds upon existing technology; the changes it introduces aren't enough to make existing products noncompetitive. Radical innovation involves large technology advancements that render existing products obsolete.

I have found the two terms that best describe the difference are "incremental innovation" and "disruptive innovation," so those are what I use. Incremental innovation delivers changes to existing products or services

---

[3] This makes sense. Process patents are usually easier for a competitor to circumvent, so rather than telling the world about them in a patent disclosure, it is often smarter to keep them as trade secrets.

to incrementally improve cost or a user's experience. Disruptive innovation delivers new products or services that are so much better than what customers currently use they fundamentally change markets. It can throw those markets into turmoil and completely rearrange the structure of the marketplace. Companies that were formerly market leaders could be left behind once the dust settles.

Much press has been devoted to the importance of disruptive innovation because large companies have not historically been very good at capitalizing on it. The literature is full of examples of market leaders falling by the wayside when they did not react to fundamental changes in technology. Western Union failed to follow the transition from telegraph to telephone. Polaroid resolutely stuck with film technology long after the world moved digital. WordPerfect ignored the transition from DOS to Windows and lost its leadership to Microsoft Word. Today's automobile industry is in the midst of a disruptive change as the world migrates from fossil-fuel-powered to electric-powered vehicles. The eventual winners are still being sorted out. In the not-too-distant future, Shell, ExxonMobil, and BP may well suffer the same fate if they do not adjust to this looming change in how automobiles are powered. Disruptive innovation rightly gets a lot of attention because the cost of missing such sea changes in markets is catastrophic to those who ignore them. Managers lie awake at night worrying about such a fate and use it as a rallying cry to stimulate innovation in their organizations.

Keep in mind, though, that incremental innovations are often more important. While disruptive innovations such as online retailing or generative artificial intelligence catch the world's attention, they occur only rarely. And when they do, the first to market is not always the one who reaps the rewards. A company that bases its future primarily on disruptive innovation is taking a long-shot bet. That is what start-ups do, and 80 percent of them fail. That is one reason large corporations often acquire rather than invent disruptive innovations. They can let the venture capital world sort out the winners from the losers.

Far more often, incremental innovations are what drive a company's results. Take Apple, for instance. In its early days, Apple pursued a strategy based largely on disruptive innovation. The Lisa computer, introduced in 1983, was the first personal computer to use a mouse and graphical user interface (GUI).[4] In an instant the fundamental way in which humans interacted with personal computers was changed. You no longer needed to type cryptic command lines to make things happen. Now you could simply point and click. This fundamental innovation had the potential to obsolete every previous computer overnight.

But it didn't. Lisa's high price ($9,995) drove away many potential users, while the lack of supporting hardware and software drove away many more. Although these problems were largely overcome with the release of the Macintosh a year later, it was too late. Far too many people had bought IBM PCs in the interim and were locked into the competition. By the time the Macintosh hit the market, most people found the cost of switching simply too high to accept. The mouse and GUI that won the market were Microsoft's, not Apple's. It was a literal example of the old maxim "The second mouse gets the cheese."

Not until the 1998 launch of the iMac, an incremental innovation in personal computers, was Apple finally able to avoid the threat of bankruptcy. They followed with two more successes, the iPod portable media player in 2001 and the iPhone smartphone in 2007. The iPod competed in a market then dominated by Sony and Philips; the iPhone in one dominated by established products such as the Treo by Palm and the Blackberry by RIM. Since neither the iPod nor the iPhone created an entirely new market, both could technically be considered incremental innovations. But Apple added considerable new value with the iTunes media library and the iPhone app store—both hallmarks of disruptive innovation. True to that definition, none of the previously dominant players are prominent in those markets today.

---

[4] The graphical user interface was developed in Xerox's Palo Alto Research Center, but Apple was the first to successfully commercialize it.

One other example of the challenge of the disruptive innovator is worth noting. Somewhat surprisingly, Kodak was an early player in the digital photography market. They introduced a low-resolution digital camera in 1973 and a megapixel-size consumer camera in 1986. But these products were only marginally successful. At the first introduction, few people were willing to spend $20,000 for a camera whose pictures did not look as good as those from a simple point-and-shoot film camera. Then, when Kodak launched their megapixel camera, there was little in the way of infrastructure to support it—no Photoshop software for picture editing, no personal computers powerful enough to edit the pictures even if the software had existed, and no photo-quality personal printers for making prints. You still had to take your digital files to the photography shop for printing. It was a business model doomed to fail.

This lack of success led Kodak to conclude the future of film was secure. By the time they acknowledged their error and took digital photography seriously, they were impossibly far behind. This kind of faulty decision-making based on the failure of an early introduction is not unique. If you don't understand why a product failed, you will not know how to predict the future of the market. As we will see in chapter 3, a similar situation at a competitor played a role in the resounding success of a business I led for Hewlett-Packard Company.

Before moving on, let's take a brief moment to review the insight we have gained from the quadrant model. First, understand that there is more than one type of innovation. Under my definition, it is classified as either product or process innovation and either incremental or disruptive innovation. The levels of investment and return vary depending on the innovation type. In general, disruptive innovation requires more investment and involves more risk but, if successful, can deliver larger returns. Incremental innovation can often be done more easily with less risk but generally delivers smaller returns. Product innovation is often the most visible element in driving a company's success—it will generate enthusiasm both with customers and in the media. Process innovation, though, can often deliver larger returns because it impacts a broad spectrum of products.[5]

3. Innovation requires commitment from all levels of the organization, starting at the top.

Lots of companies pay lip service to innovation. Their CEOs write inspiring columns in company newsletters. They sponsor company-wide "innovation contests" to send a message that innovation is important. The word "innovation" shows up prominently on their websites and in their annual reports. Like a duck hunter blasting indiscriminately into the sky in the hope that at least one shot will find a duck, these companies employ a shotgun approach on the theory that they might eventually bag a brilliant idea.

These are not the hallmarks of an innovative company. Certainly, the CEO needs to communicate that innovation is important. Innovation contests might even play a minor role. But unless you create a structured, organized plan of implementation, any innovative ideas that emerge will be all over the map. While it might be nice to implement a new system for selecting the menu items in the company cafeteria, that kind of innovation is not going to drive company growth. Innovation needs to be channeled so that what comes out of the process has a good chance of really making a difference.

4. R&D must be an innovation engine.

Most people would agree that R&D should be a key contributor to innovation. While not every innovative idea will come from R&D (innovations in marketing, manufacturing, and business processes are also essential),

---

[5] The four quadrants of innovation are not the whole story. As we will see in chapter 10, the way to manage innovation varies depending on the stage of the life cycle a market is in.

an energized R&D team with the freedom to innovate is fundamental to success.

If you ask the typical R&D engineer what it means to be innovative, you are likely to get an answer something like this: the most innovative people are the ones who come up with the largest number of creative ideas. They have the most patents. In meetings, they come up with idea after idea. They continually awe others with their creativity.

I have known many of these people throughout my career. When I query their colleagues, what I sometimes hear is "Wow, Joe is really innovative. We don't have time to explore a tenth of the ideas he comes up with. I'm glad we don't have more like him, or we'd never get anything done!"

Wishing for fewer innovative people is not the right answer. It might be better to rethink your concept of what it means to be innovative. By now you should recognize that coming up with a creative idea is only the first step in innovation. That is what is known as *ideation*, not innovation. Joe's brilliant ideas mean nothing if they cannot be commercialized. The true measure of innovation is not merely the number of ideas a person throws out or the number of patents they have been granted, but how much they contribute to the success of the business.

So how does this distinction play out in real life? Some years ago, I was assigned the task of reviewing a large portion of my company's patent portfolio to see just how well we were using our intellectual property. The task was challenging, and it was not always easy to tell whether older patents had ever found their way into products. But one thing was clear: somewhere around half the company's patents had been filed, granted, and put on the shelf, never to be heard from again. Not used in a product. Not licensed or sold to others. Not enforced against others or used as a defense when others tried to assert patents against us. The time, energy, and expense to obtain these patents had simply been wasted. It was a wake-up call that forced the company to rethink and improve its whole intellectual property strategy.

If the ideas Joe comes up with and the patents he receives are truly helping the business, he should be generously rewarded. If they sit quietly on the shelf, it is time to provide him a little gentle coaching on how to become an even stronger contributor. Most people, especially the creative ones, really want to make a difference to the business. If they clearly understand the objective, they will put all their creative energy into achieving it.

It is management's responsibility to make this happen. This involves more than just sponsoring contests and handing out awards for random innovative ideas. That's the easy part, but it's the less useful part. The real question is "What else does management do?" Do they educate people on business strategy—the target markets, the product road map, the competitive landscape? Do they fund training opportunities to help develop talented people in all departments? Have obstacles to productivity been removed so designers can focus on design work, not busywork? Are people allowed to spend a portion of their time exploring new ideas beyond their current assignments? Do interesting but risky ideas occasionally get funded? Are at least portions of those investments protected over the ups and downs of a business cycle, or are they jerked away at the first sign of a downturn? These are the hard parts, but they are the more important parts.

Throughout this book, we will learn what it takes to make your organization an innovation engine, and we will explore the various pitfalls along the way.

5. Innovation is a team game; R&D cannot do it all on its own.

Just as an army's success in battle depends on its logistics and supply lines, an R&D team's inventions will fail if they are not successfully marketed and manufactured. I am always amazed when management's response to an economic downturn is "Cut the marketing budget and lay off all the support staff so we can leave R&D untouched. That way we will be ready with a slew of new products when the recession ends." Such a reaction indicates that management does not truly understand what it takes to make an R&D department successful.

Think about it. How do you know what to invent if you are not getting feedback on what the market needs? Inevitably, one of two things will happen. You will either miss out on this market feedback and invent the wrong product, or you will send the R&D engineers out to customers to get that feedback themselves. In that case, your R&D team is doing the marketing job instead of the R&D job. It would have been smarter to retain at least a few marketing professionals who know how to do this kind of customer research.

Similarly, disproportionate elimination of such positions as administrative assistants or manufacturing engineers is false economy. The reason you have these people is that they are making important contributions to the organization. (If not, why wait for an economic downturn to get rid of them?) Lay off the administrative assistants, and your engineers now must spend time ordering parts, scheduling meetings, shipping materials to subcontractors, and performing countless other nonengineering tasks. You did not eliminate the work; you just transferred it to more highly paid staff who now have less time to get their real jobs done. A smart leadership team knows they must understand the organization's total needs and size all parts of it correctly, regardless of the phase of the economic cycle the company happens to be in.

CONCLUSION

Innovation is a key element of any successful business strategy. But as we have seen, it takes more than innovative ideas to create a winning organization. You need to turn those ideas into profitable products and services, on time and on budget. Flawed execution has doomed many more companies than lack of innovative ideas. This is a central lesson we will revisit throughout the book: innovation is not a buzzword, but rather a multiphase process that needs to be actively shaped and wielded. As such, it cannot be left to execute on its own once an idea has been introduced. How exactly does this activation play out? The next chapter explores how to turn innovation into a competitive weapon.

Key Takeaways for the Innovative Organization

1. Innovation is the ability to see opportunity in places others do not and turn that vision into profitable reality.
2. The five key tenets of innovation are:
   - Any company that wants to grow needs to be good at innovation.
   - There are various types of innovation, and the best companies excel at all of them.
   - Innovation requires commitment from all levels of the organization, starting at the top.
   - The R&D function must be an innovation engine.
   - Innovation is a team game; R&D cannot do it all on their own.
3. Innovation can be classified into four types: disruptive or incremental innovation and product or process innovation.
4. Product innovation captures the headlines, but process innovation can be even more important to a company's bottom line.
5. Innovation success depends on all parts of the organization being sized appropriately for the needs.

DISCUSSION QUESTIONS

1. What are the four types of innovation, and what are the key differences?
2. Why can process innovation be even more important to a company's bottom line than product innovation? Examples?

3. Apple was the first company to introduce a personal computer with a mouse and graphical user interface. What new value did this bring to the market? Why did this not drive Apple to become the dominant manufacturer of personal computers?
4. Why is it important to keep all parts of the organization sized appropriately for the needs?
5. How important are patents to the innovation process?

Chapter 2

Innovation as a Competitive Weapon

Now that we have established a working definition of innovation and charted out its various forms, let's explore how innovation can be wielded for competitive advantage. To begin, I'll draw from a particularly challenging juncture in my innovation journey.

Near the midpoint of my high-tech career, I took on a position as R&D manager for a small product line within Hewlett-Packard Company's test and measurement business. The main reason I got the job was that the HP[6] managers then responsible for it wanted to shut it down. They saw it as a weak business in a stagnant market, running a distant second to the market leader, with no hope for improvement.

A few of us from outside the product line saw it differently. We felt that by taking the business down a different path, HP could win in a promising new market. At the time we ran a different HP business that sold related products into that new market. We were confident that by coordinating the strategies of the two businesses, we could turn things around. We

---

[6] In this case study, "HP" refers to the original Hewlett-Packard Company, not today's computer and printer version of it, HP Inc.

were encouraged by the fact that some of their engineers and first-level managers agreed with our vision.

Three of us—my marketing counterpart, our financial analyst, and I—made our case to the R&D and marketing managers of that product line. They were unmoved. These two managers had no experience with this new market and failed to see the potential in it. They had other, more pressing matters to attend to. As far as they were concerned, the only option was to terminate the product line and move all the engineers onto more important projects.

The three of us spent several months with charts and graphs and customer testimonials trying to persuade them otherwise. Finally, in desperation, we asked if we could take over the product line ourselves. Much to our surprise, they said yes. These two managers and their boss were clearly relieved to transfer this albatross to someone else so they would not have to explain to the company CEO why they were shutting it down.

So, we took it over. Within two years, helped by those engineers and first-level managers from that product line who shared our vision, we launched a new product family that took the competition by surprise, vaulted HP into market leadership, and allowed us to capture most of the growth in that new market. Over the next six years, this outcast product line—digital sampling oscilloscopes—generated nearly a billion dollars of revenue for the company.[7] It was one of the most successful growth stories in HP's test and measurement business over that entire decade. It's a story we will explore in detail in the next chapter.

How could capable, respected managers in that business fail to see such an opportunity while a trio of outsiders could realize its potential? It would be easy to say their managers just missed it, but that's not the answer. They had a well-conceived business plan, but our proposal did not fit within it. Managers must make trade-offs all the time when deciding which businesses to pursue and which projects to fund. Rarely will they

---

[7] All nonpublic company financials have been disguised for business confidentiality, although they remain in the general ballpark.

have all the data they wish they had at the time they must act. They make the best decision they can and then move on. In this case those managers put other projects ahead of ours. To their credit, they let the product line move to a new home rather than simply die away.

The problem managers face when dealing with this kind of disruptive change arises because of something I call *the curse of the corporate business model*. It is a theme that will appear repeatedly over the course of this book.

THE CURSE OF THE CORPORATE BUSINESS MODEL

Throughout history, large corporations have demonstrated one consistent business trait: they are good at pursuing growth in their mainstream businesses but terrible at capitalizing on disruptive changes in their markets. Examples abound. Smith Corona, once a dominant manufacturer of typewriters, missed the emergence of computer word processing. Lockheed failed to respond to the transition of civilian airliners from propeller-driven to jet-powered until it was too late. Department stores like Sears and JCPenney did not appreciate the importance of discount retailing led by Walmart and Target. Although each of these companies still exist, they no longer dominate those markets.

There are three main reasons why established companies have difficulty dealing with disruptive changes:

1. **Corporate metrics and reward structures do not encourage investment in new, untried ideas.** In the corporate world, managers are rewarded for delivering continual, predictable growth and profits. This is what Wall Street investors demand, and woe to the senior management team that does not do this. Entering a new business comes with considerable uncertainty, and this does not align with the need to deliver predictable results. This same logic percolates all the way down the management chain.[8]

---

[8] For those Wall Street types who claim that stockholders do indeed factor a company's long-term investment strategy into their decision to buy or hold a

2. **Managers do not see the rewards as being worth the risk.**
   If a start-up company is wildly successful, its employees can become instant multimillionaires. That incentive can be extremely inspiring even though the chance of it happening is remote. In a large corporation, the manager who guides a new business to spectacular growth may be rewarded with a few hundred stock options or the now popular restricted stock units, and perhaps a promotion. But if it doesn't work out, their career path will likely be permanently derailed. Many managers don't see the reward as being worth the risk, especially since there are rarely any penalties for avoiding it.

3. **Managers in large corporations are inherently more conservative than their counterparts in start-ups.** The people most comfortable taking risks are not at the large companies; they are at the start-ups. Even if those people began their careers in large corporations, they soon fled what they felt was a risk-averse, overly bureaucratic culture. Managers in large companies got there in part because they liked the idea of a regular paycheck and predictable job hours. While they might be willing to take calculated risks within their present businesses, stretching outside this comfort zone becomes difficult.

   Some managers in the corporate world may object to this categorization and claim they are not shy of taking risks. If so, ask them whether they are ready to lose their job if their next risky decision does not work out. If their response is "How do you expect me to take risks if I may lose my job for a decision that doesn't work out?" simply point out that this is what happens all the time in the world of start-ups.

---

stock, note that according to Reuters, the average investor today holds a stock for only 5.5 months—just long enough to care about reaping a return from short-term profits (Chatterjee 2020).

This is why, even when CEOs try to encourage their management teams to accept more risk, they usually are not that successful. Risk-averse attitudes are so ingrained in a typical corporate culture it is virtually impossible to change them. Rather than trying to do so, the better approach is to develop ways to be successful within those constraints.

In the example at the beginning of this chapter, here is how it worked. The financial analysis we put together showed that by going after this new market, we might grow the digital sampling oscilloscope business to $20 million per year within a few years. This was not attractive to the current managers, whose existing business was already delivering over $200 million per year. In their eyes, there were less risky alternatives that could return that much or more within businesses they already understood.

For my team, the picture was much different. Ours was a $5 million per year product line, so adding $20 million per year in revenue would quintuple our business. There was no question that this was by far our most attractive opportunity. Transferring the product line to us was the right answer for both teams, a fact underscored by its eventual success far beyond our initial projections.

The fact that our business even existed within a company the size of HP might seem surprising. A $5 million business is well below the threshold most large multinational corporations would consider a viable minimum. At the time, HP divisions were designed for a minimum business size of $100 million per year, and some were five times that size.

Fortunately, Dick Anderson, the HP vice president then in charge of test and measurement, understood the curse of the corporate business model. Three years earlier he had explicitly created our organization as an "in-house start-up" to capitalize on a new market opportunity. Although we initially focused on a different market, he gave us the freedom to explore other ideas. Without his support for this atypical business structure, it is doubtful HP could have figured out how to make it work.

Today, as more corporate executives read books on innovation and recognize the curse of the corporate business model, things can sometimes swing too far the other way. Now you'll occasionally encounter an executive who says, "Why do we keep spending money inventing new products that just replace our old ones? That is not a recipe for growth. I want our investments to be in new products for new markets. That is how we will grow."

That logic may seem reasonable until you put more thought into it. Sure, you want growth, and sure, you can't get that just by refreshing your current product line. Growth in new markets certainly should be part of your strategy. But the only part? The largest part? Probably not. The opposite of growth is decline, and that's what you will get if you do not maintain your current customer base. Imagine what would happen if Toyota said, "Well, we have already invented the passenger car, so we don't need to do that anymore. Let's turn all our attention to spacecraft now. That is where the growth is."

This dichotomy lands squarely in the lap of the business unit's leadership team. They must craft a strategy that balances the need to serve existing customers with the need to grow new ones. Knowing how to do this is key to making innovation a competitive weapon.

At this point it may be tempting to imagine innovation as the domain of the R&D department, but that is a dangerous thought. While R&D needs to be an innovation engine, they cannot do it alone. Marketing, manufacturing, and finance all have key roles to play. When developing a business strategy, all four bring important insights to the table and should be equal partners. Unfortunately, this is seldom the case. Companies seem to fall into one of two categories: marketing-driven or R&D-driven. In either case, the other function ends up playing a subordinate role. Manufacturing tends to be subordinate in any case, and these days finance often perceives its role to be adversarial rather than collaborative.

For innovation to be a competitive weapon, this silo mentality must be broken. Rather than being either R&D-driven or marketing-driven, the business needs to be *market-driven*. A market-driven company breaks down the silos and puts the needs of the customer first. Every function in the business has a role to play. We will spend a good deal of this book exploring ways to do this.

WHAT CORPORATIONS DO WELL

If large corporations depended exclusively on new markets for growth, there would be a lot fewer of them around. Fortunately, most growth derives from within their existing markets. The latest sedan from Subaru or smartphone from Samsung does not open a new market; it sells to the existing customer base.

Where large corporations excel is in understanding their current customers, using that knowledge to create appealing new products for those customers, and cranking up their marketing and manufacturing engines to generate the demand and deliver the products more efficiently than their competitors. Some corporations have largely given up on the idea of funding the creation of new businesses internally. Instead, they focus their internal teams on their current markets and keep an eye on the world of start-ups for new markets. When they spot a small company that has already done the initial legwork, they make an acquisition. Google, Disney, and Salesforce are three companies that have been successful with a strategy of acquisitions. The trick is to figure out the right time to buy. If you do it too early, you might later discover the exciting new technology you bought does not really work or the market has not materialized as you expected. If you wait too long, you will pay a hefty premium.

This outsourcing of new business creation to the venture capital world does not sit too well with in-house R&D teams. "Why spend all that money to buy a small start-up when our team could invent a better product at a lower overall cost?" It does not help that they imagine the start-up's founders—ordinary engineers just like them—getting rich in the deal. (This happens far less often than imagined.)

I used to feel the same way, but over the years I have come to realize acquisitions have a place. While there are many important reasons a large company should fund the creation of new businesses in house, as we will soon see there are also good reasons to do some of it through acquisitions.

INCREMENTAL VS. DISRUPTIVE INNOVATION

The curse of the corporate business model plays out differently between the two major forms of innovation: incremental and disruptive. As we have learned, incremental innovation refers to innovation that improves existing products or services for current customers. The introduction of the latest Apple iPad is a good example. It may be faster and flashier than the previous model, but it does not create a new market; it sells to existing customers.

Disruptive innovation delivers new products or services to customers previously served by a less capable alternative. That first Apple iPad introduced back in 2010 was a disruptive innovation. Suddenly, a whole class of users discovered they no longer needed laptop computers; the simplicity and ease of use of the iPad better met their needs.

To be successful with disruptive innovation, you need to know how to deal with the curse of the corporate business model. As we will learn in chapter 7, disruptive innovations almost always need to be nurtured in an environment separate from the company's core business units.[9] Too often, managers of existing businesses will see disruptive innovations as running far afield from their core businesses. They will conclude that such innovations are more likely to reduce rather than improve near-term profitability, so they will want nothing to do with them. If you depend on those managers to successfully launch a disruptive innovation, you are likely to be severely disappointed. To avoid this obstacle, you must create an environment reasonably insulated from short-term financial requirements.

---

[9] I use the term "business unit" to describe an organization within a large company that is responsible for its own financial performance. Companies often use terms such as "division," "group," or "operation" to define their business units.

Incremental innovation is different. The driving factors behind incremental innovation are to either improve the experience for current customers, improve the profitability of the product line, or both. These align precisely with the priorities those business leaders should have. As we will learn in chapter 5, such innovations should nearly always be managed from within the product line they will benefit.

At this point you may ask, "How do I know whether an innovative idea is incremental or disruptive?" First, you should understand you cannot use the size of the project as the determining factor. Not every disruptive innovation is make-or-break for a company. This is especially true for a large multinational corporation with multiple product lines in a variety of markets—a disruptive innovation in one market may not have a major impact on overall company results. Nor is every incremental innovation a small project. A key question is whether the leaders of the existing business have the commitment, knowledge, and resources to launch the innovation successfully. To make that assessment, answer the following four questions and use them in the decision tree of Figure 2.1:

1. Does the innovative idea address a need for the company's existing customer base, or does it serve a new class of customers?
2. Is it an enhancement to an existing product, or will it replace that product?
3. If it will replace the existing product, will the leaders of that business see it as a welcome addition or as a threat to their business?
4. Are the leaders of the existing business willing and able to apply sufficient resources (staff, money, time) to the innovative idea without compromising their ability to deliver their expected business results?

Once you have determined which type of innovation you have, follow the processes defined in chapter 5 for incremental innovation and chapter 7 for disruptive innovation.

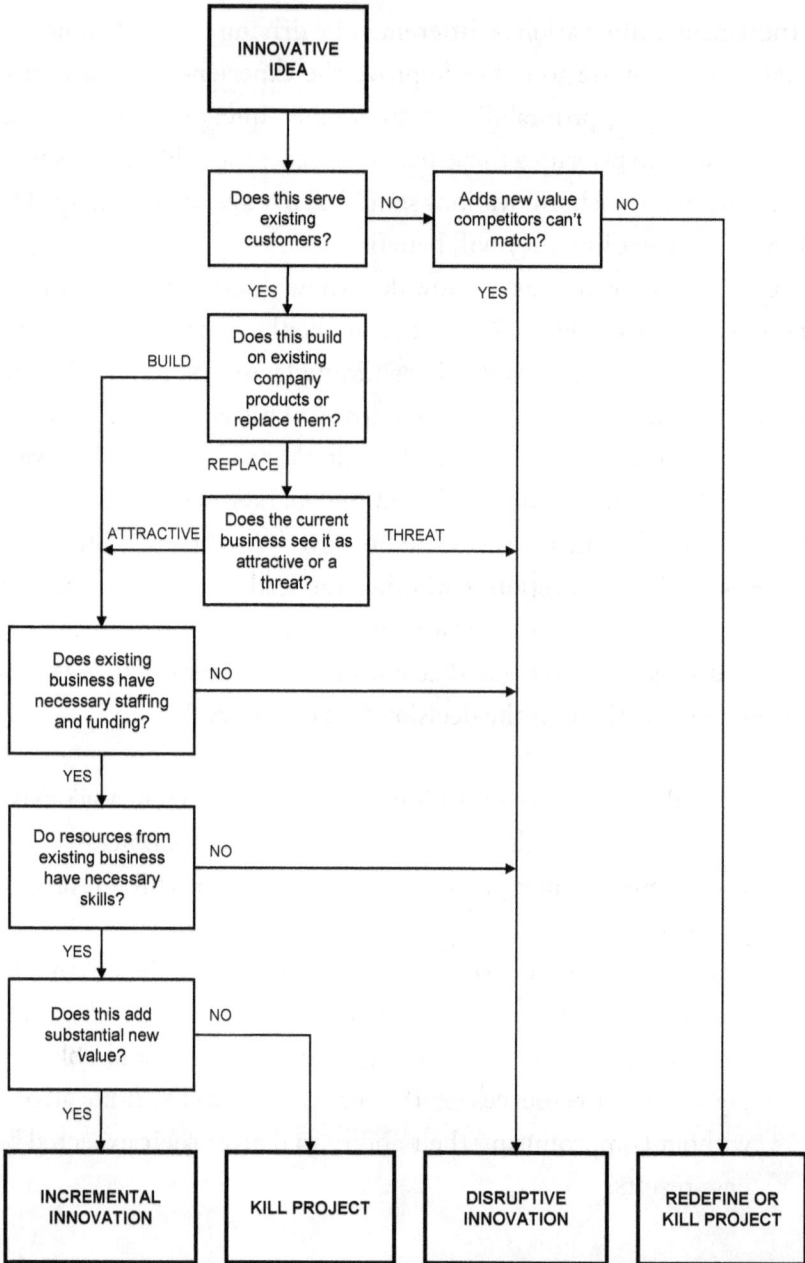

Figure 2.1: Decision tree to assess whether an
innovative idea is incremental or disruptive

TURNING INNOVATION INTO A COMPETITIVE WEAPON

What does it take to make innovation a truly competitive weapon for your organization? First, go back to the definition of innovation—the ability to see opportunity in places others do not and turn that vision into profitable reality. Many people imagine innovation to be the province of a few dedicated individuals who spend all their time walled off in quiet rooms thinking profound thoughts. It brings to mind the classic picture of the mad scientist with a brilliant idea represented by a light bulb flashing on over their head.

While it is true that innovation requires a certain amount of the "light bulb" flash of inspiration, it is not the domain of a few experts. You can do many things to foster a culture of innovation throughout an organization. In the right environment, everyone from the administrative assistant to the sales associate to the research scientist can exercise creativity and contribute innovative ideas. This will be the subject of chapter 5.

Innovative ideas are only the first step. In business, innovation is not complete until the best of those ideas are turned into profitable reality. This requires excellence in execution: selecting which of the many ideas to pursue, designing the right organization to go after them, managing the development process, connecting with customers, and measuring success. Many more businesses have failed due to poor execution than from lack of good ideas.

THE IMPORTANCE OF EMPOWERED MANAGEMENT

One other fact is important in the digital sampling oscilloscope story that opened this chapter. Its business strategy was driven not by the company CEO or his direct staff but by a team of midlevel managers familiar with the business. The CEO would have been hard pressed to describe the product line, let alone its business strategy, as the story was unfolding.

This should not come as a surprise. The CEO of a $50 billion company is much like a military general. The general can craft the strategy, but it is up to the troops in the field to turn that strategy into battlefield success. More often than not, it is the action on the front lines, not the brilliance of the strategy, that determines the final outcome. In a similar sense, it is largely within the ranks of midlevel management that the fate of a corporation is sealed.

This is at odds with the impression many people have about how businesses are run. If you listen to the business press, you are likely to conclude that the success of a large multinational company is due primarily to the brilliance of its senior leadership team. Of course, the CEO and the other "C-suite" officers—the COO, CTO, CMO, CRO, CFO, CIO, CISO, and anyone else with "Chief" in their job title—are vitally important. But in a company with thousands or tens of thousands of employees, there is only so much a team that size can directly influence. They can set the company's overall mission and direction, but it falls to the multitude of midlevel managers across all its businesses to turn that vision into reality. How well those managers carry out that task makes the difference between the company's success and failure. This is another key to making innovation a competitive weapon: creating a culture in which local management feels empowered to make decisions without waiting for CEO approval.

I was one of those managers for over twenty years at Hewlett-Packard and its spin-off, Agilent Technologies. Through all that time, from my first engineering management assignment until I became the general manager of a $100-million-per-year business, I can assure you that it was me, not the company CEO, that my staff looked to for guidance. Whenever any new directive came down from corporate headquarters, the first thing we would do is get together to decide, "Okay, how will this affect us?" A good senior-level manager can come up with quite a bit of creativity in interpreting what might seem to the uninitiated as an inflexible directive.

Here's an example. During the Great Recession of 2007–2008, I served as general manager of Agilent's digital sampling oscilloscope business. Business results throughout Agilent plummeted, and CEO Bill Sullivan took bold steps to curtail losses. One was an edict to "stop all travel." I dutifully canceled several trips my marketing department had scheduled and stopped my own travel plans. But we soon got word from our sales team in Japan that several customers were unhappy with the performance of a product we had recently introduced. They were threatening to cancel all future orders.

Relationships are especially important to the Japanese, and I did not feel we could solve the problem through emails and phone calls. Previously, I would have sent two people—a marketing engineer and an R&D engineer—to visit the customers, understand the problem, and offer appropriate apologies. I still felt a live visit was essential, but given the travel restrictions, I sent only one person, the senior R&D engineer, who knew the product better than anyone. The trip was a success, and he solved the customers' problems. The customers were impressed that I had considered them important enough to send an expert to visit them in person during a worldwide recession.

A few weeks later, Bill Sullivan came through on a routine visit and stopped by to see me. After giving him an update on the state of our business, I shared the story with him and apologized for sending an engineer to Japan in the face of a travel freeze. I explained I felt it was a critical step to save the business. Bill's response was what I would expect from a good CEO. "No need to apologize. I have to make a firm black-and-white statement because the board would have my neck if our expenses didn't show a substantial reduction from last quarter. But I expect my senior leaders to be sensible enough to do what is right for their business while staying within the spirit of the objective."

The higher you go in management, the more this kind of high-contrast thinking is necessary. You need to set a clear tone for expectations—stop all hiring, stop all travel, stop all purchasing. Then you have to depend on the managers of all the individual businesses to implement things sensibly.

This approach drives engineers crazy. "How can our CEO expect us to stop all purchasing? Doesn't he know it will kill the new product he is clamoring for us to introduce?" It is a prime reason why employee satisfaction surveys generally rank an employee's immediate management staff higher than top management. (Of course, there are always situations where people rate their immediate supervisor as "incompetent," but they often rate top management even lower for not having the sense to remove the culprit.) A good senior-level manager needs to know how to navigate the business through choppy waters successfully.

It can be a lot different in a small, family-run business. There, the founders may well control all aspects of success. In the corporate world, though, it is unusual for the CEO to have a similar level of influence. Steve Jobs of Apple was a rare exception—the CEO of a large public company still in firm control of all aspects of the business. He could do it because he was also a founder, and his imprint was large throughout the corporation. For most companies, that approach is neither healthy nor practical. A large corporation needs to develop broad bench strength so its fortunes are not inextricably tied to a single individual. Witness how the stock of another large company, Tesla, rises and falls with the latest posts on X from Elon Musk.

THE SEVEN STEPS TO INNOVATION SUCCESS

For innovation to be a true competitive weapon—one that can help you vault ahead of your competitors and dominate your markets—you need to do the following:

1. **Create an innovative environment.** Innovation blossoms when people are inspired. Chapter 5 explores how to make this happen for incremental innovation. Chapter 7 does the same for disruptive innovation.
2. **Choose the right markets.** This is the role of strategic planning: identify and segment your markets, select the best of those markets to pursue, validate your vision with customers, and create a compelling message for the team. It is equally important for both incremental and disruptive innovation. Some companies place this responsibility with R&D, others with marketing or a

separate strategic planning organization, but regardless of where it reports, it must be led by people who thrive in the quest to understand markets and opportunities. This will be covered in chapter 9.

3. **Pick the right product/service strategy.** Once you have selected your target markets, you need to decide how to go after them. What does the competitive landscape look like? What value can you bring to the market that will differentiate you from everyone else? What products or services are necessary to deliver that value? This will be another topic of chapter 9.

4. **Design the organization.** How do you design the organization to encourage innovation? What are the roles of senior management, midlevel management, first-level managers, and individual contributors? How much time should an organization allow for employees to pursue innovative ideas? What resources should be provided? How do you reward employees for pursuing innovative ideas, even if those ideas do not ultimately end up successful? We will explore these questions in chapter 5.

5. **Manage execution.** Even the best strategies will fail if not well executed. A delay of only three months in a product's introduction can have a devastating impact on its eventual success. Chapter 7 will explore methods for managing projects so that deviations can be identified early—when it is still possible to correct them.

6. **Measure success.** Success should be tracked using two different categories of metrics: financial and technical. You need to track both. This will be another topic of chapter 7.

7. **Stand out from the crowd: drive innovation by influencing the industry.** This is an often overlooked but potentially very powerful driver of innovation. You have some of the most knowledgeable people in the world in their areas of expertise. Customers want to buy from the best. Make sure they know you are the best. We will explore this topic in chapter 11.

WHERE WE GO NEXT

It is fine to talk about innovation in theory, but the best lessons come from dissecting real examples. The next two chapters do that. We will first dive into the details of the digital sampling oscilloscope story to learn how innovation can be successful in a Fortune 100 company. Then, we will explore how a company in the SMB market has innovated successfully against much larger competitors.

As you read through these case studies, understand that no example can ever exactly match the specific challenges you face in your own business. But avoid the temptation to dismiss them as irrelevant. Instead, compare how the actions in these case studies map into to the five tenets of innovation in chapter 1 and the seven steps to innovation success in this chapter. Then you will be ready to start analyzing your own situation to identify what needs to change. Subsequent chapters will explore how to do that in detail.

Key Takeaways to Innovation as a Competitive Weapon

1. Large corporations are good at pursuing growth in their mainstream businesses but terrible at capitalizing on disruptive changes in their markets.
2. The curse of the corporate business model explains why established companies have difficulty dealing with innovation:
   - Corporate metrics and reward structures do not encourage investment in new, untried ideas.
   - Managers do not see the rewards being worth the risk.
   - Managers in large corporations are inherently more conservative than their counterparts in start-ups.
3. Disruptive innovation delivers new products or services to customers previously served by a less capable alternative and should be managed in an environment separate from the company's core business units.

4. Incremental innovation either improves the experience for current customers, improves the profitability of the product line, or both. It should be managed from within the product line it will benefit.

5. Local management should be empowered through the concept of high-contrast management to have flexibility to do what is right for their businesses.

6. The seven steps to innovation success are as follows:
   - Create an innovative environment.
   - Choose the right markets.
   - Pick the right product/service strategy.
   - Design the organization.
   - Manage execution.
   - Measure success.
   - Win customers by influencing the industry.

DISCUSSION QUESTIONS

1. In the digital sampling oscilloscope case study, the previous managers of the business had planned to shut the product line down but instead allowed it to move to a new organization in the company. How difficult of a decision do you believe that was for those managers? Do you think something similar would be allowed to happen in your own company?

2. The curse of the corporate business model describes why many companies do not excel at disruptive innovation. Do you agree with the three points of this premise? Based on your own experiences, which of the three would you consider the most serious problem?

3. When deciding whether an innovation is incremental or disruptive, one factor is whether the managers of the existing business would see it as a threat or a welcome addition. How important do you think that factor is to the decision?

4. C-suite executives need to give middle managers enough author-
   ity to make important decisions without having to obtain approval
   from higher management, but they also need to protect the overall
   success of the business. What are your thoughts on how they can
   balance these two potentially conflicting expectations?

Chapter 3

Fortune 100 Case Study: Product Innovation at Hewlett-Packard Company

O ur first detailed case study includes all the elements of a Hollywood blockbuster: an innovative idea, a well-defined market opportunity, a reluctant set of senior executives, a team of innovators who refused to be dissuaded, a story of intrigue to outwit the industry-leading competitor, a fight to overcome a last-minute betrayal, and eventual success beyond the most optimistic expectations. But unlike Hollywood fiction, everything in this case study *actually happened*.

SETTING THE STAGE

Ask the average person on the street to name companies in the high-tech industry and you are likely to hear such household names as Microsoft, Intel, Apple, Google, and Dell. Who hasn't used a Dell personal computer running Microsoft Windows on an Intel microprocessor to do a web search on Google while listening to music on an Apple iPhone? It is the prototypical image of a tech-savvy consumer.

To make it all work, though, these industry giants rely on a vast network of lesser-known technology companies to perform everything from initial product design to final product manufacturing. One vitally important part of the market is an obscure $40 billion annual business known as test and measurement.

What is this business? Well, it is not enough to design a product on paper. You need to make sure it works in real life. Just as your automobile mechanic uses specialized diagnostic equipment to troubleshoot problems with your car, designers of electronic products use a vast array of test equipment to analyze the performance of their designs.

Hewlett-Packard Company started life in 1939 as a manufacturer of electronic test and measurement equipment. Books have been written about how Bill Hewlett and Dave Packard drove the original HP to become the dominant player in this business, with a reputation for high-quality products at equally high prices. If you could afford to buy test equipment from HP, that's what you did. As more than one customer told me over the years, "No one ever got fired for buying HP."

Starting in the late 1960s, HP built on this expertise to become a mainstay in the fast-growing computer and printer business. By the year 2000, that business had grown so large that test equipment was little more than a distracting $10 billion per year sideline. In that year, HP spun out its test and measurement business as a separate company, Agilent Technologies. At the time, it was Silicon Valley's largest ever IPO. In 2014, HP's original test and measurement business was spun out yet again and is today known as Keysight Technologies.

The original HP had such a powerful reputation with engineers that, in most of its product lines, it faced little competition. Oscilloscopes, though, were one of the few businesses the company had never been able to dominate. An oscilloscope is a vital part of an electronic design engineer's toolkit because it presents a graphical display of how the voltages inside a circuit change with time. Armed with this information, engineers can determine whether their designs are working properly and, if not, what changes they need to make.

The first oscilloscope, or "scope," had been invented as early as 1897, but it was the newly formed company Tektronix out of Beaverton,

Oregon, that in 1947 brought oscilloscopes into the mainstream with a revolutionary product that quickly vaulted the company to the top of this growing market. Bill Hewlett and Dave Packard were slow to react and even encouraged Howard Vollum, Tektronix's president, in his venture. Eventually, though, Bill and Dave realized HP could never truly dominate the test equipment business without dominating oscilloscopes. They set up a division in Colorado Springs, Colorado, to do just that. Over the next five decades, HP launched a multitude of oscilloscopes into the market but was never able to overtake Tektronix. It was a frustration that grated on the minds of the two company founders for as long as they lived.[10]

MARKET OPPORTUNITY

Near the turn of the twenty-first century, the worldwide market for oscilloscopes exceeded $1 billion annually. The three largest players were Tektronix, Hewlett-Packard, and LeCroy, all US companies. Although none of these companies disclosed sales data, industry analysts estimated that Tektronix held about 40 percent share, followed by HP at 25 percent and LeCroy at 20 percent. The remaining share was split across many smaller players.

Like any large market, the oscilloscope market could be divided into several segments. The largest was known as *digital real-time oscilloscopes*, or just "real-time scopes." At the time, it accounted for about two-thirds of the overall market. Designers used real-time scopes across a broad range of industries, from computers to consumer products to the highest-performance military applications.

A second segment, *digital sampling oscilloscopes* ("sampling scopes"), was a different story. A sampling scope is designed to measure much-higher-frequency signals than can be measured with a real-time scope. While at

---

[10] I still remember trying to avoid using inferior HP scopes in my college engineering classes in favor of much better ones from Tektronix. But it was HP, not Tektronix, that offered me my first job.

the time real-time scopes performed best at signal frequencies below 10 gigahertz (GHz), a sampling scope could measure frequencies as high as 50 GHz or more. (A gigahertz is a frequency of one billion cycles per second.) To put this in perspective, the microprocessor in a typical personal computer operated at around 3 GHz, well within the measurement capability of a real-time scope. Cable TV and telecommunications systems operated from 10 GHz to over 40 GHz, where the wide bandwidth of a sampling scope was essential.

Price was also a factor. While the cost of a 10 GHz real-time scope was over $100,000, a sampling scope with five times more bandwidth cost half as much. As computer, consumer, and communications systems continued to increase in speed and decrease in price, this became increasingly important. Companies could not afford to spend hundreds of thousands of dollars on test equipment to help them design products that might sell for $20 to $50 each.

To achieve this very wide bandwidth, a sampling scope had to make certain trade-offs in its design. While these trade-offs limited its general usefulness, customers who needed to measure very-high-speed signals learned to accept them. Because of these limitations, sampling scopes accounted for less than 10 percent of the overall oscilloscope market.

EMERGING MARKET: FIBER-OPTIC TELECOMMUNICATIONS

When I took over as R&D manager for HP's sampling scope product line and moved it from Colorado Springs to Santa Rosa, it was running a distant second to Tektronix. My marketing counterpart, Charlie Schaffer, estimated that Tektronix owned 75 percent of the sampling scope market and we had 25 percent. While the previous managers of the business saw little future for it and wanted to shut it down, Charlie and I had a different opinion. We saw considerable opportunity in a newly emerging market known as fiber-optic telecommunications.

At the time, long-distance telephone calls were carried by either microwave or satellite transmission systems. These systems were far from ideal. Microwave systems were sensitive to rain and snow, which caused cross-country calls to fade in and out. Satellite systems suffered from annoying delays. It could take over a second after you spoke before the other party heard you, making two-way communication difficult.

Throughout the 1990s, the telecommunications industry put considerable effort into developing fiber-optic transmission systems as a better alternative. Fiber-optic systems were attractive because they were insensitive to weather conditions and did not experience the irritating delays of a satellite system.

A fiber-optic system uses a laser transmitter to convert the electrical signals from a caller's telephone into bursts of light, then sends that light many miles to its destination down a long, thin glass fiber cable about the diameter of a human hair. At the cable's other end, an optical receiver converts the light back into an electrical signal and sends it to the recipient's telephone.

Although initially developed to carry telephone calls, fiber-optic systems have proven ideal for transporting the vast amounts of data required by internet and cable TV systems. Even cellular phone systems use fiber optics to transport calls between the towers of the system's base stations.

It would be prohibitively expensive to use a separate fiber-optic cable for each telephone call, so fiber-optic systems combine thousands of calls together and carry them simultaneously on a single fiber. The engineers who design the systems need to know what these combined signals look like so they can be sure the system is performing optimally. An oscilloscope would be the ideal tool to use, except for one problem: an oscilloscope is designed to look at electrical signals, not optical beams of light.

HP had the answer. Engineers from our central research laboratory, HP Labs, had already developed a product called a high-speed photodetector. It was a small component about the size of a cigar that did a wonderful job of converting the light from the fiber-optic transmitter into an equivalent electrical signal that could be viewed on an oscilloscope.

This photodetector had been Charlie's and my entry into the world of fiber optics. It was a low-cost accessory that did not fit with the business model of HP's oscilloscope division, so we had taken it on as part of a small product line we were then managing. We had been selling handfuls of photodetectors every month to fiber-optic engineers desperate to make their measurements.

Our connection to the fiber-optic market led us to discover an important fact. Real-time scopes could not display the rapidly varying signals from a laser transmitter, so these designers were heavy users of sampling scopes. Unfortunately for HP, most of the sampling scopes they bought came from Tektronix.

We discovered that both the Tektronix and the HP scopes could display similar pictures of the signals coming out of a laser transmitter. In fact, the HP scope was more accurate. The main reason engineers bought the Tektronix scope was for its better software—it could perform automatic tests to verify whether the transmitter's performance conformed to emerging industry standards. With the HP scope, you had to make those tests manually—a time-consuming, error-prone process. About the only reason engineers would even consider the HP scope was because we had the better photodetector.

The response from HP's oscilloscope marketing department in Colorado Springs had been to promote the higher accuracy of their product: "Why would you want to buy from Tektronix when the HP scope is more accurate?" It was a message that fell on deaf ears. Like so many marketing departments before them, HP's learned that better performance does not always lead to more sales. If a competitive product has enough performance to meet the need, customers will base their decision on other criteria. The same story has happened time and again throughout industry: Betamax video recorders may have been better than VHS recorders and Mac may be better than Windows, but neither one dominated their markets.

Charlie and I had a vested interest in this situation because the more sampling scopes HP's Colorado Springs division sold, the more photodetectors we would sell. We also saw a glaring unmet need. With both the HP and the Tektronix oscilloscopes, the photodetector was a separate component the customer had to manually attach to the front of the scope. This meant the measurement was not fully calibrated, so you could not accurately test to the true requirements of the emerging industry standards. We were confident HP could solve this problem by doing three things: build the photodetector into the scope, add the technology necessary to deliver calibrated measurements, and include the software to do automatic testing. If Tektronix did not respond quickly, we had the potential to win all the business.

This was the vision Charlie and I, along with our financial analyst Lincoln Turner, had unsuccessfully presented to the leaders of the oscilloscope business in Colorado Springs. As explained in chapter 2, these managers had no experience with fiber optics and felt they could do better by introducing improved products for their current customer base. It was a classic case of the curse of the corporate business model.

We didn't give up, and, finally worn down by our persistence, they agreed to transfer the product line to us. We would take responsibility for inventing a new product to replace HP's existing sampling scope (already a seven-year-old product), and all the revenues from this new scope would come to our business. They hedged their bets, though, by keeping the revenue from the current sampling scope for themselves and assigning a small engineering team from their organization to help us design the new one.

BUSINESS ASSESSMENT

Once we got the product charter, the first thing Charlie, Lincoln, and I needed to do was assess the situation. Using a variant of HP's strategic planning process, we developed the following insights:

1. **Market assessment.** Our best opportunity was clearly fiber optics. The only other sizeable market was in personal computers,

but at the time, most of that market had not yet reached speeds that needed the measurement power of a sampling scope. Fiber optics was the only obvious market that did. It was clearly a growing market, although its rate of growth was not clear to us. Fiber optics, however, was not a single, monolithic market. We analyzed the market and divided it into five major segments, each having different requirements (see Table 3.1). Enterprise networks connected computers within a business to its internal network. Access networks connected a home or business to the service provider. Metro networks connected customers in a city together. Long-distance networks connected cities together, and undersea networks connected continents together.

Of the five segments, the metro and long-distance networks were the most attractive. They both already used fiber optics and were moving away from earlier proprietary transmission systems toward a common industry standard. In the US, this standard was known as SONET (Synchronous Optical NETwork). In Europe and Asia, a nearly identical standard was called SDH (Synchronous Digital Hierarchy). Both standards were still under development, which meant we were not yet late to the game. Although the market was small, the potential for growth looked good. So we decided to target SONET/SDH as our initial market opportunity.

2. **Customer analysis.** We knew who the key customers were—large companies like Nortel, Alcatel, and AT&T, plus a host of smaller ones, but we did not have many connections with them. They bought their sampling scopes from Tektronix, not HP. Although they sometimes bought our photodetectors, most of them were not interested in talking to us about oscilloscopes.

We needed a way to get closer to these customers. Otherwise, we ran the risk of either inventing the wrong product or inventing one that was not superior enough for customers to justify

	Enterprise	Access	Metro	Long Distance	Undersea
Relative Market Size:	Medium	High	High	High	Low
Fiber Optic Opportunity:	Low	Low	High	High	High
Governing Standards:	Ethernet	DSL/Cable modems	SONET/SDH	SONET/SDH	Proprietary
Laser Wavelength:	850 nanometers	1300 nanometers	1300 nanometers	1300/1550 nanometers	1550 nanometers
Transmission Speeds:	100Mb/s	1.5-50 Mb/s	155-622 Mb/s	622 Mb/s to 2.5 Gb/s	2.5-10 Gb/s
Market Characteristics:	Dominated by electrical Ethernet. Fiber optics won't become important for several years. This market uses a different laser wavelength than other markets. Our initial product won't serve this market.	Market currently dominated by legacy electrical standards. Fiber optic buildout to homes and businesses is still several years away.	The transition to fiber optics is already underway. Our initial product can address this market.	The transition to fiber optics is already underway. Center of market still uncertain. Could be at either 622 Mb/s or 2.5 Gb/s. Our initial product works at either data rate and also works at either laser wavelength.	Fiber optics has already been deployed. Use of proprietary systems makes it difficult to develop a single product to meet all needs.
Example Customers:	Cisco	BT&D	Nortel, AT&T, Alcatel	Nortel, AT&T, Alcatel	Tyco, NTT
Conclusions:	Monitor market and be prepared to introduce a custom product when the time is right.	Monitor market to see whether DSL or Cable Modem technology wins. Introduce a product when the time is right.	Target this market with initial offering.	Target this market with initial offering.	Meet the needs of this market by offering special versions of our initial offering. Charge customers a premium for these specials.
Priority:	4	5	2	1	3

Table 3.1: Fiber optic market segmentation assessment at the time of this case study

switching. We knew customers would keep buying from Tektronix unless HP offered them a very good reason to change. Any time you switch suppliers, you go through a learning curve. Your new sampling scope never works quite the same way as your old one did, and you can no longer call up your friends at Tektronix to get the help you need. You are not likely to make this kind of switch unless the benefits are compelling and obvious. It is similar to the reason some people buy Ford vehicles over and over again while others stick with Toyotas or Chevys. We needed to discover the things that were most frustrating to users of the Tektronix scope and solve those with our new product.

3. **Competitive analysis.** Our only competitor was Tektronix. Their sampling scope had software advantages over the existing HP sampling scope, but it was still far from the ideal solution. It fell short in two major areas:

   - Like HP's product, it was designed to measure electrical signals. Although Tektronix sold a crude photodetector that could be attached to it, this photodetector did not meet the needs of the emerging SONET industry standard.

   - Their software, although better than HP's, did not give the user much flexibility. For design engineers, this was a real problem. When you are trying to figure out why your design does not work the way it should, you want a diagnostic tool that gives you lots of flexibility in how you make your measurements.

   Tektronix had introduced an oscilloscope for the fiber-optic market a couple of years earlier, but it had been a commercial failure. After talking with customers, we soon discovered why. Tektronix did not understand the market and had made a couple of poor design choices. First, they built their design around an existing oscilloscope platform that did not have the performance necessary to measure fiber-optic signals. Second, they had not incorporated any of the measurement software those customers needed. I got the impression someone at Tektronix had said, "We don't know what customers need, so let's just put this product on the market and see what happens. It is easier than doing the market research to find out what customers really want." But it missed the mark by so much that few customers bought it and Tektronix failed to get any feedback. I was confident that after this debacle, they were not doing anything new in this area.

4. **R&D staffing and funding.** This was a challenge. Normally, a business can use the revenues from its current products to fund development of a next-generation product. But when we picked up the charter to develop the new sampling scope, we did not get the revenue stream from the existing sampling scope; it stayed with the oscilloscope product line in Colorado Springs.

So we had to fund our development using the revenue from our current business, which was nowhere near as large. We would need to do everything possible to keep expenses down. Fortunately, the managers of the oscilloscope business had committed a small team of their engineers to work with us, and these were sharp people who were excited about the project. In these pre-Zoom days, geography was a bit of a problem—they were in Colorado Springs, Colorado, while we were a thousand miles away in Santa Rosa, California—but we made it work through regular phone calls and occasional live visits.

INITIAL THREE-YEAR PLAN

Using the results of our analysis, we developed a three-year, multiprong attack to enter the market. It included a long-term strategy to introduce the next-generation sampling scope plus two short-term strategies to quickly gain market visibility and influence. The short-term strategies would position us so that when we did introduce the next-generation product, customers would already think of us as an established player. We also wanted to give our competition at Tektronix a false sense of security. And we had to add a fourth project simply to address concerns from one of our senior managers.

Our plan consisted of the following:

- Immediately begin developing the next-generation sampling scope as a collaborative project between the Santa Rosa and Colorado Springs teams. Target the first introduction to occur within two years and follow up with additional product releases as quickly as possible in subsequent years.
- Initiate a quick-turnaround software product for use with HP's existing sampling scope, to be released within six months.
- Immediately get active in the industry's SONET standards committee.
- Develop a small, low-cost alternative product simply to satisfy the concerns of one of our senior managers.

Let's explore each of these now:

Next-Generation Sampling Scope ("BudLight project"). We knew our true competitive advantage would be to introduce a fully calibrated product with a built-in photodetector and powerful new software. Based on what little customer feedback we already had, this was something the market should get excited about. Because Tektronix had already tried and failed, we doubted they were working on anything like it, and if we executed the next two elements of our strategy correctly, they would be unlikely to start.

Our strategy called for a modular platform with a variety of plug-in modules that could be configured to the specific needs of each customer. We knew that customers in different segments of the market would need different capabilities. By making it a modular system, we could market it as a platform with strength for the future. As customers' needs changed, they would only need to buy new plug-in modules, not an entirely new instrument.

We only had a small team available to do the development, and it was split between Colorado Springs and Santa Rosa. To deal with the geographic separation, we partitioned the work so that the need for frequent interaction across geographies was minimal. The Colorado team would develop the mainframe while the Santa Rosa team would develop the plug-in modules. The project managers would need to be in daily telephone communication, but the individual engineers would not need to interact nearly as often.

With such a small team, we had to do everything we could to improve engineering effectiveness. That meant borrowing much of the design from existing products and only inventing something new when absolutely necessary. We adopted an existing hardware platform that HP in Colorado Springs had recently introduced for their real-time scopes. It was already modular, and with a reasonable amount of effort, we could turn it into a sampling scope. We could also heavily leverage the firmware[11] from that platform. We took much of the electrical design directly from HP's previous sampling scope, so the main areas of work were development of the calibrated optical circuitry for the plug-in module and the new parts of the firmware to meet the requirements of the SONET standard. (Reality, of course, is more complicated than this simplified view.)

For this first project, we wanted to keep things simple, so we only developed the two plug-in modules needed by the SONET market. Our plan was to follow later with additional modules for the other markets. This strategy worked well, and within a few years, we were selling more than twenty different modules for every conceivable customer need (see Figure 3.1), but that would come later.

---

[11] Software that is embedded inside an instrument and is used to control it is known as *firmware*.

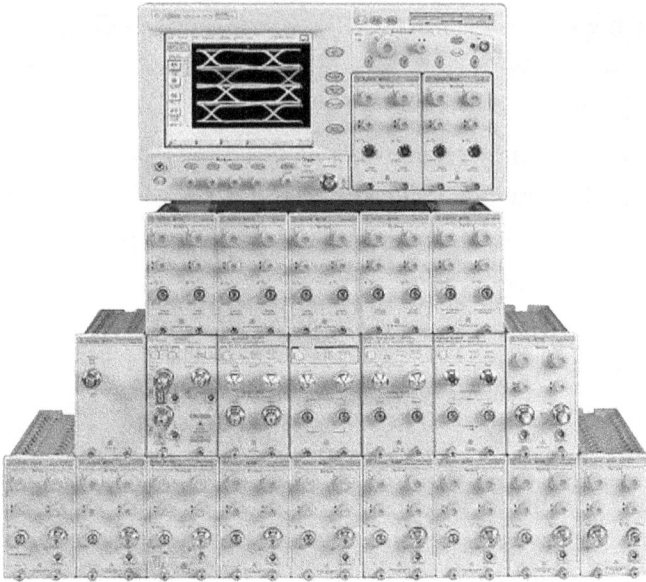

Figure 3.1: DCA product line after introduction of a full range
of plug-in modules (© Keysight Technologies Inc.; reproduced
with permission, courtesy of Keysight Technologies)

Every project needs a code name that people inside the company can refer to when talking about the project. This is essential for a couple of reasons. First, if a competitor did hear about the project, they would not be able to discern anything about it from its code name. Second, calling it "Next-Generation Sampling Oscilloscope" was an uninspired mouthful. We needed something catchier that the team could get excited about.

Our project name came about pretty much the same way as most project names do. After a full day of meetings, several of us from the Santa Rosa and Colorado Springs teams were sitting together in a bar, drinking beer. We were kicking around various potential project names based on the themes of optics or light, none of which seemed overly appealing. Suddenly, Mike Karin, the Colorado Springs project manager, held up his drink and threw out a line from a then-popular television commercial for Bud Light beer: "Give me a light!" To which we all responded, "No,

Bud Light!" And so, the name was born. Later, to mollify higher management, Mike concocted a story about it coming about as a combination of the product's ability to measure beams of light together with the fact the Colorado Springs and Santa Rosa teams were "buddies" in its development. Although the name was intended strictly for internal use and therefore shouldn't be subject to trademark laws, we did not want to take any chances. In a feeble attempt to avoid issues, we blended it into the single word "BudLight."

In a more serious vein, Charlie Schaffer strongly advocated that we market the product as something other than a "digital sampling oscilloscope." His argument was that Tektronix already owned that term in the customer's mind, and if we just copied them, customers would view us as a late entrant into a market already dominated by Tektronix. If we called it something new, we would in effect create an entirely new instrument class. Customers would then see us as the first player in a new market. Since our product would be the first to incorporate the photodetector inside the oscilloscope, it clearly broke new ground, so there was some justification for doing this.

Coming from R&D, I did not share Charlie's concern about the name. I figured customers would know it was an oscilloscope no matter what we called it. My uninspired suggestion was to just call it an "optical sampling oscilloscope." But Charlie was adamant that any name that included the word "oscilloscope" would position us in customers' minds as second to Tektronix.

Eventually, after considerable brainstorming (much of it again in bars), Charlie selected the name "Digital Communications Analyzer," or simply "DCA." We decided all our marketing and promotional material from that day forward would use this name. It would appear on the instrument's front panel. The word "oscilloscope" was banished from our vocabulary.

In retrospect, this was one of the smartest decisions we made. Just as Charlie had predicted, once the DCA was released, we quickly earned top-of-mind awareness in what customers perceived as a new market space.

They were soon using the term "DCA" in their own published technical papers, even if they took their data with a Tektronix oscilloscope. When a small start-up company launched a competing product several years later, they called it a DCA, not an oscilloscope. Even Tektronix occasionally used the term in technical papers, although never on a product. It was an early lesson for me in the power of marketing.

Quick-Turnaround Software Project ("Odin project"). This was a controversial element of our strategy, but we did it for several reasons. First, HP still had 25 percent of the market, and those customers who did own our existing sampling scope were complaining that its software was not nearly as good as what Tektronix had. While we knew a quick software fix would not close the entire gap, it would at least signal to our customers that we weren't ignoring them.

More importantly, it would give customers a reason to accept our request to talk with them. Some customers owned not only the Tektronix sampling scope but also the HP sampling scope. (They usually had many more scopes from Tektronix than from HP.) We could visit the customer, show them the software, and then talk about what they really needed. This would give us the chance to hear firsthand what they liked and disliked about both products. Later, when we introduced BudLight, we could show we had listened to them and made it a much better product thanks in part to their input.

Not everyone in HP thought this software project was a good idea. Looking at it from a purely financial perspective, it was not attractive. We planned to sell the software for about $2,000 per copy, and we estimated we might sell a handful of copies in total. At that rate, the total revenue we would generate could never pay back the engineering investment we put into the project, but Charlie, Lincoln, and I were adamant. This was not intended to be a stand-alone project; it was an essential way to get connected with our target customers. Everything we learned would be incorporated into BudLight, so it was really a way to do market research partially funded by customers. Lincoln went to battle with the HP financial community and got them to accept the idea. We launched the software after only a few months, and it did exactly what it was intended to do. Customers welcomed our visits. They were more than willing to offer insight into what the market really needed, and we earned a little money in the process.

There was one more thing in the back of our minds. We knew Tektronix would quickly hear about this software project, and we felt that could work to our advantage. We hoped they would consider it a weak attempt to shore up HP's current sampling scope and conclude we were doing that instead of inventing something entirely new. If they did hear any rumors about BudLight, they would probably attribute them to this software project and discount the threat. We wanted them to be complacent, as it would take us two years to develop BudLight. The last thing we needed was for the engineers at Tektronix to invent their own next-generation product before we could introduce ours.

SONET Standards Committee. I joined the industry's SONET standards committee for two reasons. First, I wanted to be certain BudLight would meet the emerging SONET requirements. Without this we wouldn't have a compelling, competitive advantage, and customers wouldn't be persuaded to buy our product instead of the one from Tektronix. By being part of the committee that wrote the standard, I would have the visibility I needed to make sure we met these requirements.

My second objective was to prevent Tektronix from driving the standards in a way that would lock us out of the market. We knew they had engineers on various standards committees, and Charlie had already heard rumors that they might be trying to do just that.

As luck would have it, I joined the committee at the perfect time. One of the standards then under development was something called "Optical Eye Pattern Measurement Procedure." It was intended to be the definitive document that spelled out how to use an oscilloscope to test fiber-optic systems for compliance to the SONET standard. The chair of this committee had just stepped down, and no one had volunteered to assume the role. I saw a rare opportunity to not only make sure HP was not excluded, but to also demonstrate we were the market leader. I quickly volunteered to be the new chair of the committee and was promptly accepted.

Standards committees have a legal responsibility to write standards in a way that does not unfairly exclude competitors from participating in the market. Otherwise, the government could view the committee as an illegal collusion between a few companies whose objective was to block others from competing. Before I could assume the role of chair, I had to go through a training program with the head of the full standards organization to learn how to write standards in a way that avoided this problem.

It was a responsibility I relished. As mentioned earlier, the way fiber-optic engineers made their measurements at the time was to attach a separate photodetector to the front of an oscilloscope. This meant the system was not completely calibrated, and it led to considerable uncertainty in the measurement results. The draft standard I inherited had accepted this limitation. It was written so that measurements could be made using the combination of a sampling scope and an external photodetector. Several companies had already agreed to accept it as written.

Since BudLight would incorporate the photodetector inside the oscilloscope, it would go one step further and provide a completely calibrated measurement. As chair of the committee, I could have decided to write the standard so that only this kind of fully calibrated solution was allowed. But I did not do this, for two reasons. First, doing so would have excluded Tektronix, which could have led them to raise legal objections. Since so many companies had already accepted the limitation, I could hardly have justified it. More importantly, though, doing this would have signaled to Tektronix that HP was working on a better solution. I did not want to do anything to arouse their suspicions, so I kept the standard intact, knowing that when we launched BudLight, we would have a compelling, competitive advantage.

Sure enough, just before we announced BudLight, Tektronix introduced an improved photodetector that met the requirements of my newly released SONET standard. Theirs was a bulky, expensive instrument about the size of a microwave oven, and it connected to the sampling scope through a long cable. As I expected, it still was not a fully calibrated solution.

A month later, we introduced BudLight with its fully integrated and completely calibrated photodetector. During customer visits I could tout that I knew it met the requirements of the SONET standard because I was the one who wrote the standard. When customers compared the two solutions, it was no contest: Tektronix's bulky, expensive, less accurate system or HP's sleek, economical, and more accurate solution. HP had not just made a small incremental improvement; we had made a major leap forward. It was the breakthrough necessary for customers to justify making the switch. Within three years we went from being a distant second to the unquestioned market leader with well over 50 percent market share. Over its first six years, the DCA product line generated nearly a billion dollars of new revenue for the company.

Low-Cost Alternative to the DCA. My boss, the senior manager in R&D, was not convinced BudLight was the right project, and he pushed for me to staff a quick-turnaround, low-cost alternative designed around a lower-performing sampling scope from Colorado Springs. Charlie and I understood the fiber-optic market well enough to know such a product would not have the performance necessary to meet the needs of the vast majority of customers. This manager, though, felt it was essential to get something out quickly and see how the market reacted. I knew it would likely suffer the same fate as the optical oscilloscope that Tektronix had introduced several years earlier. Fortunately, Charlie's boss, Russell Johnson,[12] the senior manager in marketing, was a strong supporter of the

---

[12] No, not the actor who played the Professor on *Gilligan's Island*.

BudLight project. Without his advocacy it might never have been allowed to continue.

I knew I had no choice, so I assigned a single design engineer to do this distracting project. The best thing I can say about it was that it was leveraged from a project that Colorado Springs had called "Saber." It was only natural that we should give our version the code name "Light Saber," which gave us the opportunity to distribute plenty of Star Wars toys to the staff. This single engineer did an excellent job on the design, but the final product was received on the market about as we expected. Based on a platform that didn't have the performance to meet customer needs, it went obsolete after just a couple of years. But good managers understand who their customers are, and not all of them are external. Some of them are their bosses.

MIDCOURSE ADJUSTMENTS

No plan ever goes exactly as anticipated, and ours was no exception. Smart managers expect this and are ready to adjust the plan as the situation changes. Our first challenge came when we tested our first BudLight prototype. It failed to meet the requirements of the SONET standard, and there were no obvious fixes. Things were looking bleak for the entire project.

In an emergency good engineers rise to the occasion, and ours were clever. After agonizing over this limitation for several weeks, the two lead R&D engineers, Mark Woodward and Randy King, came up with the answer. They adjusted the design in a way never previously considered and got it to meet the requirements of the standard.

As we connected with a broader set of target customers, we discovered that BudLight as initially defined would not meet all their needs. It had been envisioned as a tool to look at the beam of light immediately after it emerged from the laser transmitter. It could do this very well. But some customers wanted to look at the beam of light after it had traveled many miles through the fiber-optic cable. Neither BudLight nor the Tektronix sampling scope could do this. Only a minority of customers needed this capability, so it wasn't a fatal flaw, but we wanted to solve the problem. Doing so would give customers one more reason to justify switching from Tektronix to HP.

At the time, the usual way for HP design teams to deal with this kind of change was to add the new requirement to the list of product features and delay the project so the engineers had time to build it in. (This is known in the trade as "creeping featurism.") One of my first projects after joining HP had been to work with an R&D team that was already five years late introducing their product. It would take them another four years to complete it, all because of creeping featurism. Neither Charlie nor I wanted this to happen to BudLight.

We decided the answer was to invent one more product than initially planned, something called a "Clock Recovery Receiver." We would keep this as a separate project and not delay BudLight to introduce it. All my R&D engineers were fully engaged and had no time to take on anything new. Asking for more resources was out of the question, as we were already investing far more than our revenue stream could support. Our only hope was to find an outside company we could work with to develop this new product.

We found a small company in Florida that had the skill set to do what we needed. I did not have a lot of ready cash available, so we worked out an arrangement in which they funded much of the development themselves in exchange for a larger share of the eventual profits. Since this was intended to be a tool to help us sell more DCAs rather than a high-profit item by itself, Charlie and I were satisfied with this approach. Without any engineers to spare, I served as HP's liaison for this project and even ended up being the one to do the final qualification tests of the product.[13]

---

[13] Expecting a midlevel manager in a multinational company to do vocational work is not usually a good idea but is common in the kind of start-up we were trying to emulate.

We used this company for several more years until they were acquired by a competitor, but by then we had replaced their product with one of our own design. Outside partners can be useful in helping you enter a market, but be careful about depending on them over the long run; their interests will not always align with yours.

BETRAYAL AND RESOLUTION

About six months before we were ready to announce the DCA, HP's oscilloscope business went through a series of management changes. The new general manager in Colorado Springs immediately questioned why they had ever surrendered the sampling scope charter to a small team in Santa Rosa. In his view it was still an oscilloscope and should be part of his business. He refused to honor any prior agreements between our two organizations and instructed his team to take steps to bring BudLight back to Colorado.

If BudLight had been a collaboration between two separate companies, legal documents that could be enforced in a court of law would have provided protection. But since it was an internal agreement inside HP, this kind of protection did not exist. Archived memos and meeting notes are not very useful if one side simply chooses to ignore them.

Charlie, Lincoln, Russ, and I were not about to let this go without a fight. Eventually, after numerous heated discussions, we reached a compromise. Our team would keep the DCA product line and sell it to customers in the communications markets. Colorado Springs would introduce a new version of the product targeted specifically at the computer market. They would strip out the optical photodetector and the SONET-specific firmware and market the resulting product as a classical digital sampling oscilloscope. In a nod to their Colorado locale, Mike and his team christened this new project "Coors Light."

Looking back, I have to admit the strategy worked well. The Colorado team understood the computer market better than we did, and while their product never overtook Tektronix as the market leader in classical sampling scopes, it did grow our market share. To the outside world, it appeared that HP had done a masterful job of segmenting the market with two products customized for the specific needs of their respective segments. But in truth it was simply the accidental result of an internal dispute. Sometimes luck is as important as management skill.[14]

MARKET AND COMPETITIVE REACTION

BudLight clearly caught Tektronix by surprise. At the trade show where we introduced it, they sent a steady stream of engineers and managers over to our booth. Rather than chase them away (which many companies would have done), Charlie politely asked them to come back at the end of the day when they were not interfering with real customers. Then he gave them a carefully orchestrated demo while simultaneously appearing gracious. Later we heard privately that we had caught them completely off guard. Some engineers within Tektronix had been unsuccessfully advocating for a very similar product but had been turned down based on the failure of their previous optical oscilloscope. More than one discreetly inquired as to whether there were job openings in our organization.

As powerful as BudLight was, it was not an overnight success. The 25 percent of the market who were already HP customers were enthusiastic adopters. The 75 percent who used the Tektronix sampling scope were a harder sell. Tektronix had been the market leader for half a century, and this was a difficult perception to overcome. Our first two years were especially challenging. But as we continued to introduce new plug-in modules with additional capabilities, the perception gradually changed. Within three years, we had finally achieved a sustained position of market leadership.

Our position was helped by the fact that it took Tektronix six years to come out with a competitive product. They even ran out of parts for their current sampling scope a full year before they could introduce its replacement. For a twelve-month period, HP had 100 percent market share.

---

[14] Within a couple of years, Colorado Springs recognized the inefficiency of this geographic split and returned all sampling scope business to Santa Rosa.

Even then, Charlie, Lincoln, and I were not satisfied. We staffed a major new project to develop a second-generation DCA, code named "Sierra." While Tektronix was chasing BudLight, we were preparing the next leap forward.

When Tektronix finally introduced their product, it was arguably superior to BudLight. But at nearly the same time, we launched Sierra. While the two products were similar in many ways, now the roles were reversed. HP (which by now had become Agilent Technologies) was clearly the established leader. Most customers did not find Tektronix's new product to be different enough to justify the cost of switching away from Agilent. And unlike Tektronix, we made sure Sierra was compatible with plug-in modules from BudLight. This was a deviation from the industry's previous history. In the past, whenever a company introduced a new oscilloscope, there was little effort to make it backward compatible with the product it replaced. Tektronix had followed that precedent with their new product. With Sierra, customers got a clear message that Agilent was protecting their investments better than our competitor was.

In subsequent years, Tektronix exploited occasional gaps in Agilent's (now Keysight's) DCA family to win some market niches—even market leaders cannot cover every need. Eventually, however, their product line had become so weak they gave up and closed it down. As of this writing, Keysight's DCA product line—the one its previous managers once wanted to shut down—is still in business and has contributed billions of dollars of revenue to the company.

THE STORY TODAY

In 2020, the market research firm Frost & Sullivan ranked Keysight as the industry leader in oscilloscopes, finally reaching the pinnacle Bill Hewlett and Dave Packard always wanted but never achieved. While much of that success was due to the dedication of the real-time scopes team in Colorado Springs, the DCA was the first oscilloscope market segment to achieve that honor. But as Tektronix's experience shows, market leadership cannot be taken for granted. Even being the leader for half a century won't help if you do not stay on top of what is happening in today's world. It is the first step toward making your business a competitive leader, and it is what we will learn about soon. But first, we will look at one more example of innovation, this one from the small and midsize business segment.

Key Takeaways for Hewlett-Packard Company, Case Study

1. Invest the time and money necessary to obtain a clear understanding of the needs of the specific market you are targeting.

2. Develop short-term strategies that give target companies who are not currently your customers reasons to discuss their future needs with you.

3. If you are not the market leader, target a narrow segment of the market and develop a product that materially exceeds the leader's performance. If it merely duplicates their performance, customers will have little reason to switch.

4. For businesses that are driven by industry standards or regulatory agencies, take an active role in influencing those standards to assure your solution meets or exceeds requirements better than your competitors do.

5. If you are not the market leader, position your product in the minds of customers as the first in a new class of products, not as a late entry into a class already dominated by a competitor.

6. Success depends on having strong support from senior management.

DISCUSSION QUESTIONS

1. Would you classify the DCA as an incremental or a disruptive innovation?

2. How important to the overall success of the DCA do you consider each of these short-term strategies (rank them in priority 1–4)?
   a. Quick-turnaround software project
   b. Membership in industry standards committee
   c. Low-cost alternative to DCA
   d. Contract with an outside company to add one more product to the project
3. This project was conducted in a time before today's collaborative tools (Zoom, Teams, cloud-based storage, etc.) were readily available. How would you propose such a distributed-resource project be managed today?
4. What might Tektronix have done to better protect their market leadership in anticipation of a competitor introducing a disruptive product?
5. What is your opinion of the team's decision to develop a second-generation DCA that would simply replace the current product without solid evidence that Tektronix was working on a competitive product?
6. One major contribution to the project's success was the author's decision to serve as committee chairman for the SONET standards committee. He allowed the standard to adopt a less accurate approach that was supported by other companies even though he knew HP would soon introduce a much better product that would dominate the market. Discuss the ethics of this decision.
7. The project team contracted with an outside company in Florida to develop one part of the initial solution. That company was eventually acquired by a competitor. What does this suggest about the wisdom of using outside partners for this kind of service? Why do you think HP let a competitor buy the company rather than purchase it themselves?
8. HP could have decided to trademark the name "Digital Communications Analyzer," but they decided not to. Why do you think they did not trademark the name?

Chapter 4

SMB Case Study: Process Innovation at Carbon Systems

In the last chapter, we saw innovation at work in the setting of a Fortune 100 company. The SMB market is a different animal, and in this case study, we will look at how a small business, Carbon Systems, LLC,[15] has been able to compete successfully against industry giants. This case study shows that innovation is not the exclusive domain of multinational corporations or venture capital–funded start-ups. Privately owned small businesses can also be successful innovators.

SETTING THE STAGE

According to conventional wisdom, the market for desktop and laptop computers is mature and stagnant. Dell, HP,[16] and Lenovo ("the Big 3") own that market, and there is no room for anyone else. If you were to approach venture capitalists with a proposal to fund a new company that

---

[15] Full disclosure: I serve as an occasional consultant to Carbon Systems but take no credit for their success.

[16] In this case study, "HP" refers to today's manufacturer of computers and printers, HP Inc.

would sell computers to compete with the Big 3, you would be escorted out the door so quickly you wouldn't have time to say goodbye.

Carbon Systems is a US provider of desktops, laptops, and servers that has circumvented conventional wisdom. For multiple years, the company has grown much faster in their target market than any of the Big 3.

I have known David Cook, CEO of Carbon Systems, since 2010, when we both worked at TeamLogic IT. He is a tall, energetic leader in his early forties who is not only a technology expert but also a customer-focused advocate who engenders trust from his clients. His multiple decades of experience span a range of IT service providers. He has provided IT support for clients all the way from single-person businesses to large multinational corporations—including one stint at the White House in Washington, DC. This background has made him a premier expert on the needs of small businesses and the IT companies who support them. It has also given him uncommon insight into the most important unmet needs of those businesses—knowledge that has been critical to Carbon Systems' success.

TARGET MARKET: SMALL AND MIDSIZE BUSINESSES

The research firm Gartner (www.gartner.com) defines "small" businesses as companies with up to one hundred employees and no more than $50 million in annual revenue. "Midsize" businesses are companies with between one hundred and one thousand employees and up to $1 billion in annual revenue. Typical small businesses[17] include such diverse categories as attorneys, certified public accountants, doctors, dentists, retail shops, construction companies, auto dealers, hotels and inns, wineries, breweries, technology companies, and dozens of other locally owned businesses.

---

[17] To avoid overwhelming the reader with three-letter acronyms, I will use the term "small businesses" to refer to the entire SMB market.

Small businesses are heavy users of computer technology. Many of their employees use company-issued desktops or laptops, and sometimes both. Their IT networks are often elaborate, including both on-site and cloud-based servers and storage. In the past, a typical small business might hire a single IT technician to support their network infrastructure. As networks have become more complex and the threat of cybersecurity breaches more prevalent, this approach no longer works. It is nearly impossible for a single technician, no matter how skilled, to stay on top of all aspects of network design and security.

Today, many small businesses contract with dedicated IT companies to manage their networks. Such companies are known as managed services providers (MSPs). An MSP has a staff of technicians with a wide range of skills. MSPs use specialized software tools to continuously monitor a client's network and promptly address problems. With this breadth of expertise, an MSP can manage a client's network more securely and at a lower overall cost than any single technician. Three examples of MSPs with nationwide presence in North America are TeamLogic IT, The 20 MSP, and CMIT Solutions.

IDENTIFYING THE MARKET OPPORTUNITY

Many small business clients rely on their MSP to be their trusted IT advisor. When they need a new laptop, desktop, or server, they depend on the MSP to quote them the best solution. After the client approves the quote, the MSP purchases the product from one of the Big 3 manufacturers and resells it to the client at a small profit. In theory it seems simple, but in reality it is challenging. Well before he founded Carbon Systems, Dave had identified four problems that were especially troubling with this approach:

1. **It was not easy for an MSP to decide which computers to sell to their clients.** Clients expected the MSP to quote the best solutions for their needs, but that was not easy. Dell, HP, and Lenovo sold dozens of variations of laptops and desktops, and they did

not clearly define which of the many choices would be best for any particular client. Some products were more suited for large, enterprise-class customers, others for the small business market, and others for home users. Although most MSPs were authorized resellers of computers and servers from the Big 3 and might sell hundreds of thousands of dollars' worth of products every year, that wasn't enough to get much attention from any of the manufacturers. Unless the MSP sold tens of millions of dollars' worth of their products annually, none of the manufacturers would provide more than minimal support or more than minor discounts. Without solid guidance, end clients would see advertisements for inexpensive computers from their local big-box retailer and think they could just get those. They did not understand that those computers were designed for home users—they lacked the necessary business software and used less expensive components that did not have the reliability needed for full-time business use.

2. **When quoting a solution, the MSP was in competition with the manufacturer's direct sales team.** The MSP built their quote based on the price quote they received from the manufacturer. Clients would often take the MSP's quote to the manufacturer's direct sales team and get an even better price. It was frustrating— the MSP would have done all the work to research the need and decide the best solution, then Dell, HP, or Lenovo would sweep in to take the sale.

3. **When the client did buy a new computer, it took hours of the MSP's time to prepare it for use.** It was the MSP's job to get the new computer ready for the client. They would first perform the initial setup, then run Windows updates to install the dozens of patches Microsoft had released since the computer had been manufactured. They then had to remove all the advertising software the manufacturer had included to tempt the user to buy things like games or services (software known in the IT business as "bloatware"). Finally,

they installed all the other software the client needed: applications like Microsoft Office, Google Chrome, and Adobe Reader. All told, it could take as long as two hours to do this initial preparation—time the MSP's technician was not using for more productive work. The cost of this labor had to be built into the client's quote.

4. **Obtaining warranty support was a time-consuming challenge.** If a client's computer failed, the MSP would be the one who had to work with the manufacturer to get it fixed. The problem was that the manufacturer treated them no differently than they treated nontechnical end users. The MSP's technician first had to call the manufacturer's main support number and spend an hour or more waiting on hold before reaching a first-line support technician. That technician would go through a mundane script the manufacturer had created for use with the general public: "You first need to reboot the computer. Are you sure the network cable is plugged in? Is the power supply connected?" The MSP got no credit for being an IT expert who was more knowledgeable than the manufacturer's first-level support technician and had already gone well beyond that level of troubleshooting.

Eventually the manufacturer's support technician would agree that the computer needed to be repaired. Depending on the level of support the client had purchased, this could mean waiting several days for a service technician to arrive on-site. If the client had only purchased the bare-bones level of support, the MSP would have to ship the computer to a repair center in a faraway city and wait a couple of weeks for it to be returned. In either case, it would come back with factory-default settings, so the MSP would need to go through another process to get it ready for the client. While it was out for repair, the MSP would have to scramble to figure out how to keep the client operational. Most clients didn't see the need to invest in a spare computer that could quickly be put into service to cover such an eventuality, so this was a challenging job.

Dave was especially frustrated by the Big 3's failure to understand the needs of MSPs, whom they should have realized were their most important connections into the small business market. In 2017 he decided to leave TeamLogic IT and do something about it. The company he formed was Carbon Systems.

BUSINESS STRATEGY INNOVATION

Dave's first step was to decide how he could win in a mature market dominated by several multinational corporations. In their classic book *Marketing Warfare*,[18] Al Ries and Jack Trout describe the attack strategy a lesser player in the market should use when going after the market leader. Two of their principles for offensive warfare are listed below:

1. Find a weakness in the leader's strength and attack at that point.
2. Launch the attack on as narrow a front as possible.

While Dave had not yet read that book, he intuitively understood that strategy. His first step was to define a narrow line of attack. Because he understood the small-business market, he chose that as Carbon Systems' target market. He did not try to capture the entire small-business market, only the segment whose customers were served by MSPs. That would allow him to deploy strategies to address the specific unmet needs of that segment and be difficult for the leaders to match. He built his vision around four concepts:

- **Use MSPs as a virtual sales force.** By making it easy and profitable for an MSP to work with Carbon Systems, they would be encouraged to promote these products to their clients.

---

[18] I strongly advise anyone reading this book to also read *Marketing Warfare*.

- **Do not sell directly to end users.** This would reduce an MSP's frustration because they would not have to worry about clients getting better deals by going directly to Carbon Systems.
- **Make warranty support much easier and quicker.** If a product had a problem, Carbon Systems would know the MSP had already done a good job of troubleshooting. There would be no need to go through another round of entry-level diagnostics. Carbon Systems could take immediate action to fix the problem.
- **Make quoting much easier.** Rather than offering a broad portfolio of products intended to serve many markets, Carbon Systems would offer a limited set chosen to best meet the needs of small businesses.

The next question was how to source the computer hardware. Carbon Systems was too small to hire an engineering team to invent its own line of computers. The market for white label computers—ones built by local technicians from standard components ordered online—had dried up well before the turn of the century, when Dell and HP emerged as dominant.

Dave knew there were options, and he decided his best choice was to work directly with Intel. Although primarily known for their line of microprocessors for Windows computers, Intel had also created their own family of small-form-factor Windows desktops. They called it their "NUC" (Next Unit of Computing) product line. These computers were well suited for the needs of small businesses. They came in three levels of performance that covered the range most small businesses needed. By selling computers designed and built by Intel, Carbon Systems could avoid the stigma of white label computers perceived as having questionable reliability. They could promote the fact their computers were designed and built by the same company whose microprocessors dominated the market. Intel would also benefit. Sales of their NUC product line lagged far behind the Big 3, and Carbon Systems had a business strategy that could make a difference.

Figure 4.1: Carbon Systems NUC small form factor
Windows computer (photo by the author)

Intel did not offer laptop computers, so Dave decided to further narrow Carbon Systems' attack surface by starting only with desktops. Laptops and servers would come later. This kept the investment level to one he could fund internally. There would be no need to seek outside funding.

PROCESS INNOVATION

Workstations sold by Carbon Systems are similar to those sold by the Big 3. All four companies offer small-form-factor computers with similar performance at similar prices. How Carbon Systems has distinguished itself is through processes that deliver competitive advantages the Big 3 have not been able to duplicate. This capitalizes on the first principle of offensive warfare: finding a weakness in the leader's strength.

Carbon Systems has deployed four innovative strategies that help MSPs avoid the problems they face when dealing with the Big 3:

- **Make it easy for MSPs to know which computers to quote.** By limiting the breadth of solutions to the few choices best suited for most small businesses, MSPs have the guidance they need to make the right decisions.
- **Make it easy for MSPs to close deals.** Selling only through MSPs makes this easier. Clients cannot go directly to Carbon Systems to get better prices. Dave also knew his products needed to be priced competitively, so Carbon Systems aggressively negotiated with Intel's distributors to get the best possible prices.
- **Make it easy for MSPs to deliver fully configured computers.** This is a key innovation. The objective is to improve the MSP's profitability by eliminating the two hours of work they would typically spend getting the computer configured for the client. This helps the MSP improve profits because they can assign their technicians to higher-value projects.

  Dave hired a designer to write a proprietary software application that reduces configuration time from hours to minutes. Carbon Systems uses it to configure each computer before it is shipped. When placing an order, the MSP simply specifies how they need the computer configured and what software applications to install. Carbon Systems runs the installation, then ships the computer directly to the end user. The only thing the MSP needs to do is make a phone call to the client to go over how to connect the computer to the network. There is rarely a need to send a technician on-site.
- **Make it easy for MSPs to quickly deal with warranty problems.** This is perhaps the most innovative component of the strategy. Since all Carbon Systems products are sold by authorized MSPs, the MSP is the one who performs first-level troubleshooting. If they confirm the problem is real, they simply call the Carbon Systems support line and request a replacement. No questions asked, Carbon Systems ships a replacement computer

by overnight express to arrive the next business day. The MSP then returns the defective computer back to Carbon Systems. This eliminates the need for Carbon Systems to maintain a staff of field support technicians who must go out to do on-site repairs. This rapid response is a significant competitive advantage that the Big 3 have not been able to replicate.

ORGANIZATION DESIGN

For the first year, the business was small enough that Dave could manage sales and operations himself. He hired technicians to handle production. NUCs were received from Intel without Windows installed. A technician ran the software application that installed Windows, applied patches, and installed all the additional software the MSP had ordered. They then shipped the fully configured NUC directly to the end customer using expedited delivery.

Within a year, business had grown to the point Dave couldn't handle it all on his own. He assessed his strengths and determined he was best suited for business planning, sales development, and customer relationship management. Where he could use help was in managing the day-to-day production operation and vendor negotiations, especially with Intel and their distributors.

Dave found the perfect fit for director of operations in John Rosebaugh, a former client of TeamLogic IT. Dave knew John was a skilled negotiator because the two had faced off in negotiations multiple times when John was Dave's client at TeamLogic IT. John also showed the level of attention to detail necessary to keep on top of daily operations.

As the business has continued to grow, Carbon Systems has added staff in R&D, sales, and postsale customer support. Many of these resources are geographically remote from their California headquarters. They reside in such locales as Mexico, the Philippines, and the eastern United States—all locations with lower cost structures than California. John Rosebaugh, now their COO, explained to me how he and CEO Dave Cook make it work:

We have status meetings on Zoom every week. The cameras are turned on, and it almost feels like the remote workers are here or in an office just down the street. Everyone gives status updates around what they are currently working on, what is coming up, and something we call a "maybe"—something far-fetched, not formed well, but something to be explored. We also reserve time for what we call "party crashers." These are things people can't solve on their own; they need help from management. It could be something like a permissions issue or a vendor not responding. Dave or I will help them overcome that challenge. We don't try to completely steer the ship, but if we don't think the team has the ability to do something on their own, we will step in and either start it or guide it along the way. We then hand it back to them as much as possible as soon as possible. We also made sure our infrastructure is transparent to remote workers. They have their own IP phones and full access to our internal IT network. Developers can remote into the computers and servers in our system as if they were here. Cybersecurity is critical, and we have made sure the whole network is locked down.

PRODUCT LINE EVOLUTION

Carbon Systems' vision has always been to be the go-to hardware supplier for MSPs who support small-business clients. Desktop computers were only the first step in this vision. Other important product lines include laptops and servers. Since Carbon Systems is too small to staff an engineering team to invent custom products, their answer has been to initiate relationships with two manufacturers in Asia. Those manufacturers have adapted existing laptop and server product lines to meet Carbon Systems' requirements. This kind of relationship with Asian manufacturers is not unusual. The Big 3 often do the same, sometimes using the same companies used by Carbon Systems.

As with desktops, the performance and reliability of Carbon Systems' laptops and servers are similar to those from the Big 3. Carbon Systems' competitive advantage comes from applying the same superior customer support processes already in place for desktops. Again, it is an advantage the Big 3 have not been able to replicate.

RESPONDING TO BUSINESS CHANGES

In the technology world, business is rarely static. Carbon Systems faced a major challenge when Intel decided to exit the NUC business. Since Carbon Systems' advantage is based on its processes rather than its products, Dave could have decided to approach one of the Big 3 manufacturers and negotiate a deal to resell their products. But Dave had already been planning ahead. He knew it was important to nurture a close relationship with the senior executives of Intel's NUC business. Intel executives had invited him to join their customer advisory board, a small team of their most important customers who help shape their future strategy. This gave him early notice of Intel's plan, so he was able to develop a strategy for dealing with it. He placed larger orders for NUCs than he would otherwise have done, and he stayed in contact with NUC senior executives to stay aware of everything they could legally share with him.

The crisis was averted when Intel closed a deal to sell the product line to Asus. Since then, Dave has established relationships with Asus executives similar to the ones he had with Intel. Asus invited him to join the board of advisors for their NUC product line, so Carbon Systems remains well positioned for future success.

FACING THE FUTURE

Carbon Systems is a tiny company compared to Dell, HP, or Lenovo. The book *Marketing Warfare* would classify them as a guerrilla company that should follow the principles of guerrilla warfare: never try to launch a direct frontal attack on the market leaders and focus instead on a narrow segment of the market too small for the leaders to care about. Keep a lean organization that can quickly respond to change, don't add too many layers of management, and keep focused on providing improved solutions for their target market. Trying to act like a leader is a strategy destined to fail.

David Cook and John Rosebaugh understand this. They have stayed focused on delivering improved solutions to their target market while keeping a lean organization. Nothing is ever certain, but they are well poised to continue their success in the coming years.

Key Takeaways for Carbon Systems, LLC, Case Study

1. If you are not the market leader, identify a target market large enough to be profitable but narrow enough not to put you in the crosshairs of the leader.
2. Carefully define the target market and the most frustrating unmet needs of customers in that market.
3. Develop strategies that solve at least some of the customers' unmet needs in ways that the leaders are unable to replicate.
4. Do not try to take on too much initially. Start with a limited strategy and add to it as results allow.
5. As the business grows, perform an honest assessment of your strengths and weaknesses. Add talent conservatively to help fill the gaps.
6. Consider establishing processes that allow you to use geographically dispersed talent effectively.

DISCUSSION QUESTIONS

1. Would you classify this as incremental or disruptive process innovation?
2. Who do you consider to be Carbon Systems' primary customer— the MSP or the MSP's end client? What should be the strategy for their secondary customer?
3. ConnectWise, a leading supplier of software tools for MSPs, estimates that in 2019 there were approximately forty thousand MSPs

serving US small businesses. About eight thousand of those accounted for 80 percent of the business. They estimated the overall business has been growing at a rate of 13 percent per year. Using these assumptions, what is your assessment of Carbon Systems' potential market opportunity?

4. How likely do you think it will be that Dell, HP, or Lenovo will begin to consider Carbon Systems a threat to their businesses? What possible actions could the leaders take to counter Carbon Systems' advantages?

5. What actions should Carbon Systems take to continue growing its business and deal with emerging competitive threats?

6. Given that computers from Carbon Systems provide the MSP with more value than similar ones from the Big 3, how important is it for Carbon Systems to price match the Big 3?

Part II

Empowering Incremental and Disruptive Innovation

Chapter 5

Creating an Environment for Incremental Innovation

I n the book's opening section, our aim was to establish a shared understanding of the various types of innovation while demonstrating that innovation is a competitive weapon that can be wielded by businesses of all sizes. In this section we will show how managers can drive innovation in real time and real space, within everyday work environments optimized for productive change. This chapter covers incremental innovation. The next two chapters cover disruptive innovation.

DRIVING INCREMENTAL INNOVATION

It is easy to imagine that the way to foster an innovative environment is through such perks as free lunches and dinners for employees, free daycare for their dogs or children, video games or foosball tables in break areas, after-work parties, or any of a multitude of other perks for employees and their families. Since the COVID-19 pandemic, the ability to choose a hybrid work schedule that includes both in-office and work-from-home days has become popular, especially for people who have families and are unlikely to go to after-work parties.

These can all be ways to help a company attract and retain strong talent, but they are not the fundamental components of innovation. Once you have enticed that talent to join your company, you need to do the things that encourage them to innovate.

Incremental innovation deals with improving existing products or processes for existing customers; you are not trying to invent new products for new markets. The people best suited for pursuing incremental innovation are the ones who deal with the current business every day. They know it better than anyone and are in the best position to propose improvements. Marketing specialists who work closely with customers learn what those customers wish the current product did better. R&D engineers who designed that product have the best insight into how to improve it. Production workers who build it know where the bottlenecks are in the process and what can be done to improve them. All these people have a vested interest in making the product even better.

Contrast this with the approach some companies have tried in which incremental innovations are the responsibility of a separate "innovation department" staffed by people who do not have responsibility for the current product line and are not measured by quarterly profits. This does not usually work very well.[19] When the innovation department comes up with ideas for improvement and presents them to the leaders of the existing business, they are met with skepticism: "How could people who have never faced our challenges think they know our customers better than we do?" It is the classic case of local management resisting the message, "I'm from corporate, and I'm here to help."[20] Without solid support from those local leaders, the improvement ideas are doomed to fail.

Everyone in the organization has the ability to contribute to incremental innovation. The key is knowing how to encourage them to do so. There is no single formula for doing it. Different companies do it in

---

[19] Although as we will see in chapter 7, this does work well for disruptive innovation.

[20] A phrase I used regularly with Hewlett-Packard divisions when my job at corporate manufacturing was to lead them through a disruptive innovation in printed circuit technology.

different ways, but here are six things I have found helpful to encourage employees to innovate:

- Educate employees on the company's business strategy and priorities.
- Explain that innovation is more than just ideas for new products or services.
- Allow time for employees to explore innovative ideas.
- Provide employees with resources to support innovation.
- Encourage collaboration.
- Reward both success and constructive failure.

Let's explore each point in more detail:

Educate Employees on the Company's Business Strategy and Priorities

Innovation will not thrive unless employees have at least a basic understanding of the company's business—its mission, priorities, and core values—and believe they can make a difference. Without that knowledge, any useful innovations they come up with will be due to little more than blind luck. It is the responsibility of senior leadership to craft a compelling business strategy and communicate it throughout the organization.

My first exposure to the concept of business strategy came when I joined Hewlett-Packard Company. During the new-hire orientation program, we learned about HP's seven "Company Objectives." First published in 1957, they had been updated slightly over the years. Although the original Hewlett-Packard no longer exists, these remain an example of what companies today often call "core values" (Packard and Hewlett 1974).

It is interesting to note that HP's first objective, profit, rarely shows up in company core values statements today. Companies today see profit as an outcome of other core values rather than a core value itself. It is almost as if they are embarrassed to say they are in business to make a profit.

Amazon's 2009 annual report, for example, spoke with pride about how the company had 452 goals, and not one of them included the words "gross profit." But as Bill Hewlett and Dave Packard were quick to point out, they did not put profit as HP's first objective by accident. They knew that none of the other objectives could be achieved without first making a profit.

Hewlett-Packard Company Objectives

1. **Profit**: To achieve sufficient profit to finance our company growth and to provide the resources we need to achieve our other objectives.
2. **Customers**: To provide products and services of the greatest possible value to our customers, thereby gaining and holding their respect and loyalty.
3. **Fields of interest**: To enter new fields only when the ideas we have, together with our technical, manufacturing, and marketing skills, assure that we can make a needed and profitable contribution to the field.
4. **Growth**: To let our growth be limited only by our profits and our ability to develop and produce technical products that satisfy real customer needs.
5. **Our people**: To help HP people share in the company's success, which they make possible, to provide job security based on their performance, to recognize their individual achievements, and to ensure the personal satisfaction that comes from a sense of accomplishment in their work.
6. **Management**: To foster initiative and creativity by allowing the individual great freedom of action in attaining well-defined objectives.
7. **Citizenship**: To meet the obligations of good citizenship by making contributions to the community and to the institutions in our society which generate the environment in which we operate.

By no means do I suggest that HP's corporate objectives are the right ones for any other company. That is a decision for a company's senior leadership team to make. A detailed discussion of how to define a business strategy, mission and vision statements, and the company core values is beyond the scope of this book. Refer to the references section for more information.

Once senior leaders have crafted the business strategy, the next step is to assure priorities are aligned throughout the organization. Innovation won't flourish if individual department priorities don't match what the overall business is trying to accomplish.

Senior leaders must first share a version of the business strategy that is appropriate for all employees to see. Next, business-unit managers must assess how their specific strategies align with the top-level objectives and what changes may be necessary. There may be areas where a department's strategy doesn't align well with the corporate strategy. Often, it is a simple matter to make immediate adjustments. For more difficult matters (such as when a marketing department has made commitments to customers that do not align with top-level objectives), assign teams to follow up and develop plans to resolve them. Once strategies are aligned, the organization can confidently move forward with innovation.

It is just as important for employees in a small business to understand their company's core values, but the process can be simpler. The owner or company president can host the kickoff meeting to share the story. Follow-up meetings can simply be discussions between employees, with the owner checking in to get regular updates on how things are progressing and what they can do to help.

Here is an example of how I once used core values to motivate my team to achieve powerful results. In 2010 I started a position as president of the newly formed TeamLogic IT franchise in Santa Rosa, California. As an MSP our job was to provide IT support for the many small businesses in the region that depended on their computer networks but did not have IT expertise on staff. We were in competition with several established players, and, after a year in business, we were growing only slowly. I needed to build our business by orders of magnitude quickly. I was sure the opportunity was out there. Many more businesses needed IT support than were currently getting it.

At the time we were a small office with only a handful of employees. Most of those were IT technicians who were skilled at solving technical problems in a client's network but not at seeking out new business opportunities. I knew that for us to achieve our growth objectives, I would need their help.

Here is how I did it. I started by establishing a goal for the total dollar amount of growth we needed to achieve over the next six months. Just sharing that number would not be very helpful; I needed to translate it into objectives the technicians could relate to. While we could meet our goal in many ways, I decided to promote one specific approach. I set four targets for the next six months: the number of new monthly contract clients we needed to sign, the number of new computers we needed to sell, the number of new servers we needed to sell, and the number of new business phone systems we needed to sell. If we hit all those numbers, I was confident we would achieve our growth target.

I pulled everyone together over a pizza lunch to explain the overall growth objective and why it was important. Next, I shared the explicit targets I had created and some ideas for how the team could help achieve them. Finally, I laid out what would be in it for them: if we met each of the four targets before the end of the six-month window, I would buy everyone a new iPad (this was in the days before iPads were common, and none of the technicians owned one). If we missed even one of the numbers by as little as one day, I would not.

While handing out free iPads would not be cheap, I knew if we made the numbers I could well afford to do so. If we missed the goals, it would be demoralizing for everyone, but I was not too worried. I was confident we could hit them. If we failed, I would come up with a plan B that would help the team stay motivated.

To keep the goals at the top of everyone's mind, our office manager (my wife, Nicki) bought four large poster boards. She drew giant thermometers labeled from 0 to 100 percent on each of them (including a generous margin in case we were especially successful) and placed all four in the break room. As we added new clients, sold new computers, sold new servers, and sold new phone systems, she filled in the appropriate thermometer to update the percentage total we had achieved toward the objective.

I watched as nearly every day during lunch, the technicians gathered around the posters to brainstorm what they could do to close the gap. One technician knew the owner of a fifty-person ambulance company. He talked to that owner, learned they were unhappy with their current IT provider, and got us the opportunity to quote. We not only won that business, but just a couple of months later, that owner also referred us to an even larger client. Other technicians helped close deals for computers and servers at other clients.

As new clients came on board, it became clear that the current staff was getting stretched thin. If I did not get them some help, I worried they might get frustrated and stop trying to seek out new business. Although it would be a financial strain, at this point in our company's life it was far more important that we grow. If we did, I knew profitability would follow.

I posted job openings and let everyone know about them. One technician quickly referred a friend he knew from a competitor. She was a strong contributor. I also hired one more technician, which got us to the appropriate level of support. Although neither of those people had been with the company at the beginning of the challenge, I let them both know they were also eligible for the iPads. Enthusiasm remained high, and one by one the thermometers burst through their goals. Before the six months was up, we had hit every objective. Everyone got their iPad.

Largely because of this level of enthusiasm, the franchise also benefited. For two years in a row, we were honored for being the fastest-growing franchise of any in the TeamLogic IT network, and in the second year we were named franchise of the year.

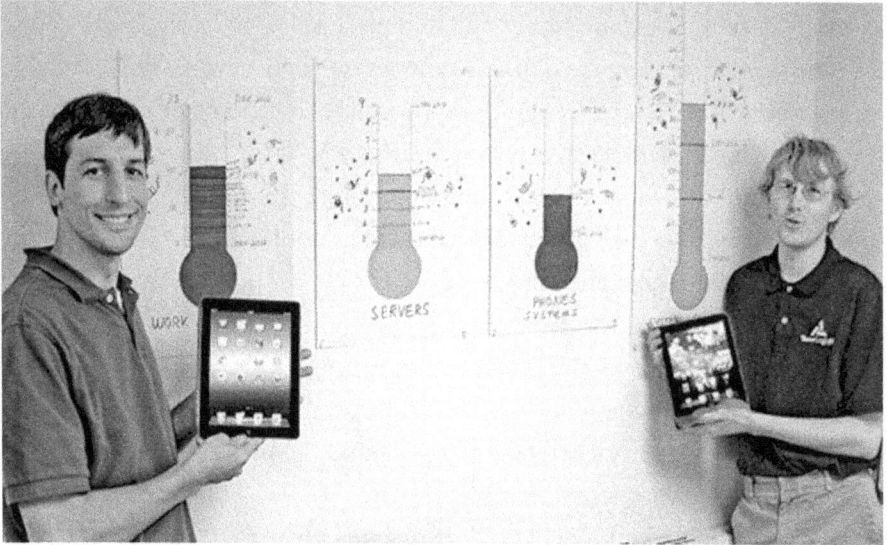

Figure 5.1: Technicians David Cook (L) and Travis Toomey with the iPads they earned by helping TeamLogic IT exceed targeted growth goals (photo by the author)

Educating employees on the company's business strategy is a key step toward encouraging innovation. By understanding the company's target markets and growth goals and the actions needed to achieve that growth, employees have the insight they need to make useful contributions.

One common response from managers is "I don't want to share information about the company's business strategy with every employee. If they leave and go to work for a competitor, I don't want them telling their new employer everything about what we are doing." But you should not be sharing strategy at such a detailed level that competitors could use it against you if they knew it. A high-level story of your mission, priorities, and competitive situation is something your competitors have undoubtedly already figured out. If not, they probably are not much of a threat.

Explain That Innovation Is More Than Just Ideas for New Products or Services

Once people understand the company's business objectives, they can be effective at innovating. Disruptive product innovations like cryptocurrency or the SpaceX spacecraft capture the public's imagination and make innovation headlines. There is more to innovation, though, than new products or services. Process innovations can be even more powerful.

Here is an example. In 2007 Amazon introduced their first e-book reader, the original Kindle. It was a disruptive product innovation that preceded Apple's iBooks on the iPad by three years. Financial analysts and the trade press latched on to it as a potentially make-or-break product for Amazon.

A few years earlier and to little media fanfare, Amazon launched two new services, Amazon Marketplace and Amazon Advantage, which were made possible by significant process innovations. They gave third-party sellers, from the smallest one-person store to the largest nationwide retailer, the ability to sell their products through Amazon—either directly or through links from their own websites—and reap a portion of the profits. These process innovations were largely ignored by the trade press.

Although Amazon does not report detailed sales figures, experts estimated that by early 2010, the Kindle reader had sold about three million units and accounted for approximately $700 million in new revenue over three years—an impressive result. But according to Amazon's 2009 annual report, revenue from third-party sales accounted for 30 percent of Amazon's unit sales in that year alone—ten times more in one year than Kindle since its inception. And Amazon said they did not care whether you bought from one of their third-party sellers or from Amazon directly; they made about the same profit in either case.[21] This is a good example of

---

[21] It is even better now. Amazon reported that in 2023, third-party sales accounted for 60 percent of sales through Amazon's store.

how process innovations can have a far greater impact on company results than any single new product introduction.

Allow Time for Employees to Explore Innovative Ideas

Innovation takes time. If you do not give employees time to innovate, they will not do it. Carving out innovation time for direct labor employees may seem challenging because they are often measured on their level of output over a full eight-hour workday. Any time spent on work that does not directly contribute to that output would appear to be a detriment to productivity. It may seem easier to carve out time for salaried employees whose productivity is not measured on a minute-by-minute basis, but that is a bad distinction. Employees who work on a production line, for example, will likely have ideas for how to improve their manufacturing processes in ways the manufacturing engineers who support that same production line will not. Giving those employees time to develop their ideas may well produce improvements whose benefits far exceed the cost of the time spent working on them.

The business's leadership team needs a vision for how much time they want to reserve for work on innovative ideas outside people's regular jobs. In my tenure as general manager for Agilent Technologies' Digital Communications Analyzer business, I targeted an average of 5 percent of the overall organization's time for blue-sky innovation—the "discovery phase," where an idea was initially being explored. When times were good, I might stretch that to 10 percent, and when times were slow, I might pare it back to 2 percent, but if at all possible, I never eliminated it entirely.

A quick mental calculation says 5 percent over a forty-hour workweek amounts to two hours per week per person. That was my benchmark for the time I would support for people to spend exploring true blue-sky ideas on their own, although I never envisioned there being a two-hour-per-week clock that tracked each person's time. They might not do any work on innovation some weeks and more than two hours in others.

I did not expect this two-hour window to be a black hole of time for employees to use in secret. My expectation was that the employee or a small group of employees would come up with the glimmer of an idea on their own, then discuss it with their direct manager to get support and any help they might need to explore it. The easiest way to do that is as part of the regular meetings a manager should be having with each of his or her employees. While these meetings (typically weekly but no less than monthly) should focus on the employee's regular job—what they have accomplished since the last meeting, what they will be doing next, and what help they may need from the manager—they should also include an opportunity for the employee to discuss any potentially innovative ideas they have.

The manager must treat the employee's idea with respect. It would be impossible in this first discussion for the employee to have already created a detailed strategy. At this point the manager just needs to give the employee permission to spend more time exploring it. The manager needs to listen carefully and ask what they can do to help.

The worst thing the manager can do is shoot an idea down by saying something like, "We tried that before, and it didn't work." Thomas Edison failed thousands of times in his attempt to invent the light bulb. If he had quit after his first attempt, we would all still be using candles in the dark. A better approach is to point out the earlier failure and encourage the innovator to learn why it failed. Sometimes, the problem is not with the idea but with the execution. And even when the idea was the problem, it is possible new thinking will uncover ways to improve it.

Managers need to make sure employees know that innovative ideas are supported. Employees need to be confident that if an idea does not work out, their career should not be damaged. Managers also need to be sensitive about how they push innovation with employees. If they try to force it by expecting a new innovative idea at every meeting, ideas will soon deteriorate into ones more intended to satisfy the manager than provide useful innovation.

Not every employee will be a fountainhead of innovation. Some, because of personal preference or lack of confidence, will prefer to focus only on what they need to do for their job. They can still be important contributors. It is the manager's responsibility to assure that employees understand what they need to do to keep their skills relevant. We will discuss the need for employee development plans shortly.

Direct labor employees and their immediate managers who are measured primarily on weekly productivity may need more encouragement to spend time on something that does not contribute directly to this week's targets. Rather than expecting those employees to explore innovative ideas on their own, it can be helpful to reserve time for discussions in group sessions. This is easiest for teams that have regular group meetings. Production teams may have a daily fifteen-minute meeting to discuss plans for the day, supplemented by a weekly meeting either at the end of the week to review how things went or at the beginning of the week to discuss plans for the coming week. Use these as opportunities to discuss roadblocks or potential improvements that could be solved through incremental innovation.

If an innovative idea proves promising, the innovators will need more time to pursue it. At this point the idea moves out of the discovery phase's 5 percent time threshold and into the realm of a formal, separately funded project. My approach has been to encourage the individual department managers in R&D, marketing, manufacturing, human resources, and finance to sponsor innovation projects within their departments. Some ideas are small enough that they can be staffed at that level. Others, either because they stretch across department lines or because they are too large for a single department to fund, would be brought to my functional staff for discussion. We would select the most promising of these to fund at the overall business level and review progress regularly—either monthly or quarterly, to address the following:

- Has the issue been fully defined? Who are the target customers, and what is their problem to be solved?

- What work has been done to confirm this is a valid problem for the target customers?
- What alternatives for solving the problem have been explored, and why does the proposed solution appear to be the best one?
- What is the current state of the project, and what are the most significant risks?
- What are the next steps?
- How can the leadership team help remove any obstacles?

My bosses, of course, still expect me to meet my quarterly targets, so it is up to me to figure out how to fund these initiatives while remaining on target. (Hint: one way to do this is during the annual planning process. Negotiate targets that won't be nearly impossible to meet, and include a "slush fund" either as a separate line item called "innovation projects" or buried within other budget line items. This allows you to fund a few initiatives outside the normal deliverables while still meeting your numbers. It also shows that top management has a true commitment to innovation, and it gives managers whose bonuses depend on meeting or beating targets an opportunity for better rewards.)

Provide Employees with Resources to Support Innovation

Resources come in many forms: people, money, tools, time, and training. The relative importance of each varies depending on the stage an innovation is at. For ideas that have already been defined and approved for investigation, success depends on ensuring that the management team has committed the necessary resources: an adequate staff of people with time and space to do the work, adequate funding, and whatever physical or software tools that may be necessary.

I learned a lesson about the importance of resource availability when I first joined Hewlett-Packard. Part of my indoctrination was to hear the many legends about how Bill Hewlett and Dave Packard had shaped their company. Innovation was a priority for them, and they did many things to encourage it. One famous story had to do with a storeroom known as "lab stock" in the R&D department. It was where engineers went to search through hundreds of storage bins to get whatever parts or tools they needed for the projects they were currently working on.

Although this storeroom was primarily intended to supply parts for company projects, Bill and Dave said engineers were welcome to use lab stock parts for personal projects too, under the theory that this could improve their skills and possibly even result in something that would find its way into an HP product. (I speak from personal experience. My knowledge of the optical technology we used in the DCA came about partly because of the experience I gained designing an optical timer from lab stock parts to measure the accuracy of the shutters on my large-format camera lenses.) Since engineers often worked after hours or on weekends, Dave had decreed that lab stock was to always remain open.

One day the lab stock manager, apparently tired of having to restock parts for what he believed were engineers' personal projects, decided to install a padlock to keep the storeroom closed after hours. One weekend, Bill Hewlett came in to work on a pet project. He went to get some tools out of lab stock, discovered the padlock, and went ballistic. He tracked down a large bolt cutter, cut off the offending obstruction, and left a note that lab stock was never to be locked again. When the manager arrived on Monday morning, he was furious until he saw the signature on the note.

Because of the possibility of theft, not every CEO would be comfortable leaving a storeroom unlocked after hours. Even Agilent Technologies, the company spun out from HP, was not immune. We had to deal with an employee who stole over a million dollars in equipment and was selling it online. (The offender was sentenced to two years in prison and ordered to pay more than $1.2 million in restitution.) The message is not that you should leave storerooms unlocked, but rather that providing employees the resources they need to be creative can be a tremendous inspiration to innovation.

In the earliest stages of innovation—the "blue-sky" phase before an idea has solidified—priorities are different. Here, one of the most important priorities should be to give employees access to knowledge that will help them in their innovation process. One way to do this is to send them through training courses relevant to the subject at hand. Even for disruptive innovations, it is rare that an idea would be so unique as to never have been explored. It is often possible to find courses targeted for professionals and put on by universities or consulting firms that provide valuable insight.

Another option is to send representatives to trade shows relevant to the topic under consideration. By carefully planning the purpose for such visits and giving each attendee specific objectives to be accomplished while there, you can gain insight into the overall state of a market, its competitive landscape, and the most important concerns of potential customers already in that market.

This is a good time to talk more generally about training. In today's rapidly evolving world, no successful company can afford to remain static. Skills that employees have today may not be sufficient a year or two down the road. Smart managers know they need to invest in employee training for those evolving needs. Some types of training are basic components of the job: teaching a new production employee how to assemble products or a new engineer how to deal with the intricacies of the engineering documentation system are necessary steps. That is not the kind of training I mean. Rather, it is training that advances an employee's skill: helping a software designer grow their understanding of generative artificial intelligence by sending them through a course on AI or improving an administrative assistant's efficiency by putting them through a course on efficient time management, for example.

My philosophy has always been that it is the employee, not the employee's manager, who should be responsible for proposing a development plan. The manager can help in several ways. First, share with the employee the kinds of expertise the company is likely to need in the future. Then encourage the employee to think about what that could mean for them. I have always tried to fund one significant development activity per year for each employee as long as it is something they have proposed, it is financially feasible, and it aligns with expected future needs.

Finally, I will point out one example of what not to do to encourage innovation. During one downturn in the economy, the finance department for the company I was at considered all sorts of ways to cut expenses. At the time, the company contracted with an outside provider for janitorial service. Among other tasks, janitors would come in each evening to clean the floors around employee workspaces and empty the wastebaskets from those workspaces.

In one brilliant move, the CFO decided to eliminate this nightly janitorial service. Employees were told that from that day forward, they would need to keep their own workspaces tidy and empty their own wastebaskets into trash bins at the ends of the building. No thought was given to simply reducing the frequency of janitorial service to every other day or even once per week. It was just eliminated. Of course, the offending CFO did not empty his own wastebasket; that was a task for his administrative assistant.

Employees recognized this as a typical example of how some managers have no idea how to manage a business successfully during a downturn. What do you think would happen if Taylor Swift was told that because of cost cutting, she would henceforth need to empty her own wastebasket? The manager who made that decision would soon be out of a job. Treat your employees like the superstars they are, not like pesky derelicts you only keep around because you have no choice.

Does the company I described still force employees to empty their own wastebaskets? I wouldn't know. I soon left to join a more enlightened company.

Encourage Collaboration

One person might come up with an innovative idea, but inputs from others will make it better. This is especially important for ideas that could improve products or services. Regardless of where the idea came from, at minimum, R&D and marketing must work together to refine it. R&D will provide perspective on technical feasibility and marketing on customer value. For innovations involving physical products, a representative from manufacturing should also be included. When appropriate, a financial liaison can help analyze the financial impact. For process innovations that impact only a few areas, a subset of people from those areas may be all that is necessary.

Avoid assigning too many people at the outset. Large committees invariably force too many compromises because they try to accommodate the full range of constraints imposed by all members of the team. Instead, limit the initial team to only those few members who are enthusiastic supporters of the idea. Once the concept has solidified, add people with relevant skills as necessary.

The case study of the Hewlett-Packard DCA showed how collaboration between teams separated geographically by a thousand miles was key to the success of an innovation project even in the days when collaboration tools were limited to phone calls, email, rudimentary file sharing, and live travel. Collaboration has become easier in the Digital Age. Tools such as Zoom or Microsoft Teams make it easy for geographically dispersed teams to meet and communicate almost as effectively as if they were in the same room. Online chat via voice, video, or text using tools such as Slack or Discord, when configured for private teams, can be a powerful supplementary tool. Cloud-based storage solutions such as Dropbox or Google Drive make it easy for teams to share files and other data. Generative AI tools such as ChatGPT or Microsoft Copilot can be useful for helping teams shape their strategies.

Time zone differences can be a challenge for geographically dispersed teams. For companies like Carbon Systems that primarily serve one geography (in their case North America), hiring staff who reside in Asia or Europe means those people will be working late at night or early in the morning in their local time zone. COO John Rosebaugh says that Carbon Systems does everything possible to accommodate those remote staff members. He particularly stresses the importance of considering them long-term members of the team who are treated no differently than local staff in California.

Reward Both Success and Constructive Failure

Employees need to be confident that working on an innovative idea is encouraged even if the idea does not pan out. People who have put dedicated effort into innovation need to be rewarded not only to recognize their own work but also to show others that innovation is important. Rewarding success is easy; rewarding failure is more of a challenge. Not every failure is equal, and not every failure deserves to be rewarded.

Here is an example of rewarding success. During my tenure as head of marketing for one Agilent Technologies division, my boss, Dave Bass, vice president and general manager of the division, wanted to encourage our R&D team to deliver an important new product on time and on budget, so he offered an incentive. If the team met the goals, he would allocate several thousand dollars they could use to improve their work environment however they wanted.

Encouraged by this incentive, the R&D team delivered the product on time and on budget. Team members considered various ways to use the reward money—things like an espresso machine, video game console, or pool table. In the end, they decided to buy a high-end massage chair and place it in the lab. Knowing their success was due not only to their own work but also to the help they received from marketing, manufacturing, and finance, they put out the word that the chair was not just for the R&D team's use; everyone in the organization was invited to use it.

Offering employees an incentive for achieving the goals is a good way to recognize their contributions, but it does not need to be something pricey. A group lunch, a small bonus, or an inexpensive memento presented at a group meeting may well be sufficient, ideally something that is related to the objective. Be sure to check with your finance department to assure that you know the tax implications of any reward you provide.

In the example above where I rewarded everyone with iPads, I made sure they all knew the iPads remained company property and would need to be surrendered if the employee left the company before the iPad was written off our books.

Deciding whether to reward failure depends on why the failure happened. Some failures occur for reasons the innovators could not have known in advance. Not every possible market opportunity will pan out, and not every technology will prove feasible, but you won't know that until you spend time investigating them. These kinds of failures should be recognized as delivering useful insight that helps shape future strategies.

Here is an example from my days as general manager of the Digital Communications Analyzer business at Agilent Technologies. My product marketing manager suggested we explore the market for digital video test equipment[22] as a possible extension of our product line. While that was not a market we currently served, it appeared we could adapt the DCA platform to do so. She proposed that before we spent time designing prototypes, we do a better job of understanding the market. She took several people to the premier event for that industry, the National Association of Broadcasters trade show in Las Vegas. Each team member was given specific objectives that would help us learn more about the market, the customers, and the current competitors.

At the trade show, the team discovered this would not be an easy market to break into. Customers needed dedicated test equipment designed expressly for the video test market. They would not be willing to pay a premium for equipment that had more capability than necessary. That meant we would not be able to modify our existing products; we would need to invent entirely new ones. The team also discovered that video production companies were low-margin businesses that were always looking for the least expensive products to do the job. This meant our profit margins would be low. Technical requirements were driven by industry standards,

---

[22] Equipment used by broadcast television stations to maintain their networks.

so the performance of our products would need to match what was already on the market. There was not an obvious opportunity to provide break-through technology at a competitive cost. The market was dominated by only two companies, and the only reasonable way to enter this market would be to acquire one of them. But given the low margins and uncertain growth rate, this did not make sense. We only needed that one weekend at a trade show to collect all the information necessary to make our decision. We shelved the idea, but to recognize the value the team had provided, I presented each of the innovators with a certificate to take their family to a nice restaurant for dinner at company expense.

Failures that are the result of avoidable problems do not deserve either public recognition or public humiliation. Often, these are due to poor performance on the part of the project manager or others on the team. Sometimes they are due to insurmountable obstacles that should have been obvious had the appropriate research been performed at the outset. These do not necessarily need to be career-limiting events for the people involved. Treat them as learning experiences by having a private discussion with each team member to help them understand how to use what they learned to improve their ability to deliver on future projects.

CONCLUSION

The six steps above demonstrate that innovation is inseparable from the collective experiences of employees—their everyday work, goals, successes, even failures. To harness the full potential of incremental innovation, managers and other stakeholders must design work environments primed for productive change.

Now that we understand the concepts behind managing incremental innovation, it is time to turn our attention to disruptive innovation. We will start with a historic case study that proves the curse of the corporate business model is not a new phenomenon. It has been with us for well over a hundred years.

Key Takeaways for Creating an Environment for Incremental Innovation

1. Incremental innovation is best performed by the people who deal with the products or processes every day, not by a separate organization with no responsibility for monthly or quarterly results.

2. Every employee in the organization can contribute to incremental innovation. It is not the domain of a chosen few.

3. The organization's leadership team is responsible for coaching employees on how to perform incremental innovation. Six key steps are listed here:
   - Educate employees on the company's vision and priorities.
   - Explain that innovation is more than just ideas for new products or services.
   - Allow time for employees to explore new ideas.
   - Provide employees with resources to support innovation.
   - Encourage collaboration.
   - Reward both success and constructive failure.

DISCUSSION QUESTIONS

1. Why is incremental innovation best performed by the people in the existing business rather than by a separate "innovation" organization?

2. This chapter proposes that an organization reserve an average of 5 percent of people's time to pursue the "blue-sky" phase of innovation. Other authors propose 10 or even 15 percent of their time. What are the advantages and disadvantages of increasing the percentage of blue-sky time?

3. How should an organization deal with an incremental innovation that appears promising but is beyond its ability to staff or fund?

4. Some managers claim to promote innovation, but their actions do not support it. What can managers do to assure that their actions align with their promises?
5. When should failure be rewarded, and when should it not be rewarded? Examples?

Chapter 6

Historical Case Study: Disruptive Innovation in the Railroad Industry

Disruptive innovation, by definition, changes a market so significantly it throws that market into turmoil. The change may be slow to develop, and at first it may only impact a minor part of the market—a part the leaders do not consider important. But if those leaders fail to act, they may eventually discover the new technology has improved to the point where it serves the needs of a broad swath of the market much better than they do.

Authors of books on innovation cite many examples of modern companies that have missed the impact of disruptive change. But that danger is not a new phenomenon; it is built into the bedrock of industry. To better understand, let's look at an example from the first half of the twentieth century: the transition of railroad locomotives from steam to diesel power. Although it is from an earlier time, the lessons it teaches are just as applicable today. Indeed, the transition of locomotives from steam powered to diesel powered is the classic story of how market leaders who are adept at deploying incremental innovations to their current product lines fail when they try to apply that same approach to disruptive innovation. As is often the case, a newcomer to the market with no attachment to a legacy product line becomes the eventual winner.[23]

The transition from steam to diesel evolved in several phases over several decades—a pace that was easy for the leaders to ignore. Like in the myth of the frog who never hops out of a pot of water as it is slowly brought to boil, steam locomotive manufacturers ignored a slowly changing landscape until it was too late. But unlike the myth of the boiling frog, this story was all too real for the participants (see Figure 6.1).

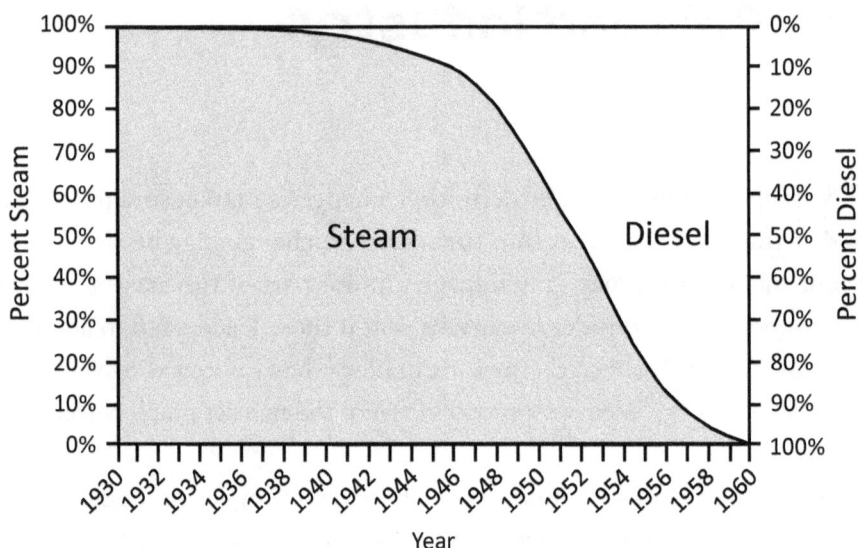

Figure 6.1: Percentage of locomotives in service by type, 1930-1960. Electric locomotives (not shown) never accounted for more than 2% of the total. Adapted from Norton, Hugh S., "The Locomotive Industry in the United States, 1920-1960, A Study in Output and Structural Changes," *The Railway and Locomotive Historical Society Bulletin*, October 1965, pp. 66-76

---

[23] For a full history of this story, see Dr. Albert J. Churella's excellent book adapted from his doctoral dissertation, *From Steam to Diesel: Managerial Customs and Organizational Capabilities in the Twentieth-Century American Locomotive Industry.* Quotes attributed to him are from this book.

A CENTURY OF STEAM

The first successful steam locomotive in the United States, the Tom Thumb, entered service in 1830. Steam-powered railroads quickly became popular, and by the time of the American Civil War, they had established themselves as the country's premier mode of transportation for both passengers and freight.

As rail transportation grew throughout the late nineteenth and early twentieth centuries, trains became longer and heavier. While heavy trains could be pulled by two or more moderately powered steam locomotives, that was not attractive because each locomotive needed its own crew. Railroads saw the solution as being larger and larger locomotives that needed only one crew, and they pressed builders to design them. As a result, the power of a typical mainline steam locomotive grew from under one thousand horsepower in the 1870s to over six thousand horsepower by the 1940s (Figure 6.2).

Steam locomotives performed best over a relatively narrow range of conditions, so the kinds of locomotives a railroad needed depended on a variety of factors. Locomotives for mountainous terrain were different from those for flatlands. Locomotives for light branch lines were smaller and lighter than those for highly trafficked main lines. Locomotives for passenger trains were faster and more streamlined than those for freight trains.

Figure 6.2: Union Pacific's Big Boy locomotive, manufactured by Alco, was the largest steam locomotive ever built, heavier than a 747 aircraft. In 2019, Union Pacific restored No. 4014 and returned it to service for select tours (photo by the author)

Because of these varying needs, locomotive builders designed their factories around three constraints:

- **Work closely with customers to design locomotives specific to their needs.** Their primary customers were officials in railroad motive-power departments. These were the people who made the final decisions on which locomotives to buy. They expected to be involved in every step of the design process to assure the final locomotive met their needs. Since each railroad had its own set of requirements, locomotives designed for one railroad were rarely ideal for use on another railroad. Even when they were, those railroad motive-power officials refused to accept locomotives they had not helped design.

- **Optimize the factory for small-batch production.** Because every railroad had its own needs, steam locomotives were never built on high-volume assembly lines like those found in the automobile industry. Factories consisted of shops dedicated to specific tasks spread out across multiple buildings. Locomotives under construction were moved from one shop to the next until they were completed.

- **Hire skilled craftsmen to deal with manufacturing limitations.** Tolerances on parts for steam locomotives were so large that those parts were never truly interchangeable. Builders hired legions of skilled craftsmen to do whatever was necessary to make things fit. This was challenging enough for the original builder; railroads also needed their own cadre of experts who could fit replacement parts onto locomotives undergoing repair or maintenance.

By the 1920s steam locomotive manufacturing was dominated by three companies: Baldwin, Lima,[24] and the American Locomotive Company (Alco). Baldwin was the oldest of the three, having built its first steam locomotive in 1831. Lima began building locomotives in 1881, starting with small, gear-driven

---

[24] Pronounced "LYE-ma," not "LEE-ma."

locomotives for the logging industry. Alco was formed from the merger of eight smaller builders in 1901 to combat Baldwin's market dominance.

In the first decades of the twentieth century Alco and Baldwin regularly traded leadership positions, while Lima remained a distant third. According to Dr. Churella, in the 1920s Alco averaged 47 percent share of the market, Baldwin 39 percent, and Lima 14 percent.

FIRST THREAT TO STEAM: ELECTRIC-POWERED LOCOMOTIVES

The first threat to the dominance of steam came in the 1890s. Steam locomotives generated clouds of smoke that left quantities of soot across the landscape, on passengers, and on any clothes hung out to dry by families living nearby. Serious accidents had occurred in smoke-filled tunnels because engineers were unable to see dangers ahead. Crews and passengers alike found it difficult to breathe while traversing long tunnels, occasionally with fatal results.

In response to public outcry, railroads worked with a small number of builders to develop electric-powered rail vehicles as an alternative to steam. General Electric and Westinghouse were the two most successful entrants in this field. GE introduced their first lines of electric streetcars and steeplecab electric locomotives in the 1890s. Westinghouse followed a similar timeline, introducing their first boxcab electric switch engine in 1895 (Figure 6.3).

Figure 6.3: GE steeplecab electric locomotive (left);
Westinghouse boxcab electric locomotive (right)

Alco and Baldwin could have used this as a way to gain expertise in this new technology. The opportunity was certainly there. GE contracted with Alco for the metal frames (known as "carbodies") in their electric switchers and Westinghouse with Baldwin for theirs, but neither Alco nor Baldwin went further than that. Both left the design of the electric motors (known as "traction motors"), the associated electrical circuitry, locomotive final assembly, and all sales and technical support to GE and Westinghouse. The locomotives were sold as GE or Westinghouse products with scant mention of Alco's or Baldwin's contributions as subcontractors.

Railroads found the expense of installing the necessary overhead wire (known as a catenary) or third rail to be overwhelming. They concluded they could afford to deploy electric locomotives only along a few heavily traveled main lines, primarily in the eastern United States. This limited the market opportunity and convinced both Alco and Baldwin it was not important to their future.

SECOND THREAT TO STEAM: RAILCARS

Although in the United States, fully electric locomotives never became a serious alternative to steam, the idea of using electric traction motors to drive the wheels remained appealing. Electric locomotives were more efficient and needed less maintenance than steam engines, and what maintenance they did need could be performed by a smaller number of employees whose skills differed from those of traditional shop workers. The question was how to provide the electricity without installing external power lines.

Gasoline engines had proven successful in automobiles, so in the early 1900s, several companies explored using them in the railroad industry. The best gasoline engines of the time could produce only a few hundred horsepower, which was not enough to compete with even a small steam engine. But they had enough power to drive a single railroad passenger car, which led to the development of a new class of rail transport: a self-contained, single-car passenger unit known as a railcar.[25] While a few used the gasoline engine

---

[25] Colloquially known as a "doodlebug."

to power a mechanical transmission, most used it to power a generator that drove traction motors similar to those on streetcars. By being completely self-contained, a railcar could operate over lightly used branch lines that could never have justified the expense of a catenary. Again, GE was an early entrant into the railcar market, with Westinghouse following a few years later.

Both GE and Westinghouse exited the railcar business during the First World War to focus on selling only the electrical subassemblies. When demand for railcars resumed after the war, it was a new company, Electro-Motive Corporation (EMC), that emerged as the leader.

Led by the dynamic Harold L. Hamilton, EMC began in 1922 as little more than a marketing company. Hamilton had been an engineer for Southern Pacific, a manager at the Florida East Coast Railroad, and later a marketing executive for White Automobile Company. He saw railcars as a way to combine his knowledge of railroads with his knowledge of marketing and high-volume automobile manufacturing to establish a new niche in the railcar market. Rather than following the traditional steam-era approach of building railcars that were unique for each customer, EMC sold standard designs that could be built economically in large volumes on assembly lines similar to those in the automobile industry (see Figure 6.4).

Figure 6.4: First railcar from Electro-Motive Corporation, 1924

Hamilton soon found that officials in railroad motive-power depart-ments resisted the idea of purchasing railcars of a standard design. They were used to working closely with steam locomotive manufacturers on custom designs and did not want to surrender that role. But Hamilton was undeterred. He simply went over their heads directly to top manage-ment. His message was that EMC's gasoline-powered railcars were more economical to purchase and operate than custom designs—a message that resonated with company presidents and financial officers far more effec-tively than with motive-power officials.

This was the start of something steam locomotive manufacturers had never before worried about: marketing. They had worked with the same railroads for so long that everyone knew each other intimately. EMC did not have that luxury, so Hamilton applied concepts he had learned selling automobiles to gain a foothold in the railroad industry. Besides marketing to a new class of railroad executives, EMC offered warranties, training, and stocks of spare parts—things steam locomotive manufacturers had never done before and would never do as well as EMC.[26] In later years, after introducing its first diesel passenger locomotives, EMC took its story to the general public through displays at trade shows and the 1939 New York World's Fair. Passengers were soon clamoring for travel on fast, mod-ern, smoke-free trains.

With no manufacturing facility of its own, EMC purchased gasoline engines from Winton Engine Company and electrical components from GE. They contracted with several companies to build the final cars. By 1930 EMC had sold over five hundred railcars and captured nearly 90 percent of the railcar market. All three steam locomotive builders were content to leave this small market to the newcomer.

---

[26] Builders of steam locomotives rarely kept spare parts. If a locomotive under repair needed a replacement part, the railroad would contact the builder and have them make a new one from existing blueprints or templates. The locomotive could be out of service for weeks.

THIRD THREAT TO STEAM: THE FIRST
DIESEL-POWERED LOCOMOTIVES

In 1923 New York City adopted the Kaufman Act, which prohibited the use of steam locomotives anywhere in the city beginning in 1926, although after legal wrangling the date was delayed by five years. While railroads had already converted many of their main lines in the city to electricity, they still used numerous small steam locomotives for switching duties. These would need to be replaced, and soon.

Because of its high flammability, gasoline was never an attractive option for switch engines. Several railcar accidents had resulted in fiery infernos that killed dozens of people. Diesel fuel was much less flammable[27] and offered other advantages. Its 20 percent improvement in energy per gallon over gasoline reduced fueling expense and, even more importantly, reduced the cost of transporting that fuel. According to Dr. Churella, one tank car of diesel fuel could replace eight hopper cars of coal. And diesel engines had no need for water to generate steam, which made them especially attractive to railroads serving the arid West.

The Kaufman Act drove both Alco and Baldwin to make their first forays into technology for small diesel[28] switch engines—not because they believed the time for diesel locomotives had arrived, but because they were pressured to do so by railroad officials. Alco's first attempts were in collaboration with GE and Ingersoll-Rand, but once again, Alco's only contribution was the carbody (see Figure 6.5). The locomotive was marketed by Ingersoll-Rand, and the few produced by this partnership were largely unsuccessful outside of New York City.

---

[27] Locomotive diesel fuel has a flash point above 125 degrees Fahrenheit, compared to gasoline's flash point of −49 degrees Fahrenheit.

[28] Technically "diesel-electric" because the diesel engine drives a generator that powers the electric traction motors.

Figure 6.5: America's first diesel switch engine, a boxcab design
from a collaboration between Alco, GE, and Ingersoll-
Rand, introduced in 1924 (builder's photo)

Electro-Motive Corporation had struggled at the beginning of the
Great Depression because the market for railcars was saturated. A savior
came about in an unlikely way. General Motors, at the time the world's
largest automobile manufacturer, wanted to expand its capabilities in
diesel engines for motor vehicles. GM purchased the Winton Engine
Company in 1930 and discovered EMC was Winton's largest customer.
So GM bought EMC too, retaining Harold Hamilton as its president. In
1941, GM renamed it Electro-Motive Division (EMD).

GM had no interest in the diesel locomotive market. It had purchased EMC merely to assure that Winton would retain its best customer. But Harold Hamilton had his own vision of the opportunity.
He knew that railroads across the nation were using EMC's railcars
in a way he never intended. Rather than keeping them as single-unit
passenger cars, they were disconnecting steam locomotives from

their smaller passenger trains and using the railcars as underpowered locomotives.

Hamilton saw this as an opportunity to develop the passenger locomotives that railroads really needed. He knew the gasoline-powered technology of railcars would never do that; it needed to be diesel. He worked with Charles F. Kettering, GM's head of research, in an attempt to convince CEO Alfred P. Sloan to invest in diesel locomotive research. Sloan was unimpressed, but when he instructed them to shut down that research, Kettering's response was to tell Sloan he needed $500,000 more to move it forward. After additional pressure from Hamilton, Sloan relented and gave Kettering the money.

Meanwhile, Ralph Budd, president of the Chicago, Burlington and Quincy Railroad, was in the market for a lightweight, internal-combustion-powered passenger train that could be run profitably over CB&Q's many lightly traveled branch lines. He reached out to Edward G. Budd (claimed by some sources to be a distant relative), owner of the Edward G. Budd Manufacturing Company, a producer of stainless-steel automobiles and railcars. Ralph Budd placed an order for a three-car trainset consisting of a streamlined locomotive and two permanently connected passenger cars. Edward Budd accepted the order and began the design even without knowing how it would be powered.

Harold Hamilton got wind of the order and immediately approached Edward Budd with a proposal to power the train with a 660-horsepower Winton diesel engine. Budd agreed and designed it in. The CB&Q Railroad introduced this first diesel-powered trainset as the *Burlington Zephyr* (Figure 6.6). On its initial run, it covered the thousand miles from Denver to Chicago in just over thirteen hours—half the time of the previous record. It introduced the public to the benefits of diesel technology and even got Hollywood's attention—the *Zephyr* was the top-billed star in the 1934 movie *The Silver Streak*.

THE MOST ILLUSTRIOUS RAILROAD TRAIN IN THE WORLD
Built of stainless steel—Electric shot welded—Rides on articulated trucks. Powered by an eight-cylinder, two-cycle, 660 horse-power, oil-burning Diesel engine. Runs on roller-bearings—Air-conditioned—Equipped for radio reception

Figure 6.6: Postcard introducing the Burlington Zephyr,
1934 (author's collection)

The market for passenger trainsets was limited, so in 1935 EMC followed with a line of 600- and 900-horsepower switch engines. Railroads found them economical to operate and soon stopped buying steam switchers. Both Alco and Baldwin realized they needed to compete, so both accelerated their diesel programs. As market leaders, they did not need to be first to market with the new technology. All they needed to do was introduce diesel switch engines that were roughly equivalent to those from EMC. Most customers would prefer to buy from a company they already knew and trusted. But neither did that.

Alco had introduced its first in-house-designed diesel switch engines even before EMC, but it was the result of underfunded development by a top management culture that was more interested in paying dividends to shareholders than investing in research. Their diesel switchers were built in the same low-volume production shops as steam engines, with none of the economies of scale. Using processes and labor more suited for steam, the resulting locomotives were expensive and unreliable. According to Dr. Churella, several years later the Pennsylvania Railroad determined it cost them three times more per mile to maintain Alco locomotives than EMC units.

Baldwin was even less successful. The Great Depression had forced it to declare bankruptcy in 1935, and even after reorganization it was left with little money to fund development. Baldwin's first diesels used engines that had been designed for stationary service by a company with no experience in the railroad market. When placed in locomotives, they vibrated severely and were so large that locomotive engineers had difficulty seeing the track ahead during switching service.[29]

As with Alco, Baldwin's switch engines were built on shop floors designed for steam by people who had no experience with diesels. Baldwin's president, Samuel Vauclain, remained firmly committed to steam and surrounded himself only with like-minded executives. With no champion inside the company to advocate for them, Baldwin's diesel switch engines never won more than a tiny share of the market.

THE END OF STEAM

In 1939, EMC (soon to be renamed EMD) introduced the first reliable diesel locomotive suitable for mainline freight service, the groundbreaking F-series of streamliners. The initial model shown in Figure 6.7, known as the FT, introduced railroads to the concept of multiple-unit diesel freight locomotives. While one FT locomotive produced only 1,350 horsepower, four tied together provided a total of 5,400 horsepower—enough to compete with the best steam locomotives. Even if one unit failed, the remaining three had enough power to keep the train

---

[29] Baldwin had acquired the Whitcomb Locomotive Works, a builder of small diesel locomotives, in 1930, but somewhat surprisingly kept them separate from the rest of the company and didn't even attempt to use Whitcomb as the source of the diesel engines they needed for switch engines.

Figure 6.7: EMC's FT locomotive was the diesel that heralded the end of steam. It was designed as multiple units of 1,350 hp each that could be controlled by a single crew. The four Great Northern units in this photo produced a total of 5400 horsepower (Archive PL/Alamy Stock Photo)

moving, if a bit slower. And unlike steam, the entire four-unit consist could be controlled by a single crew.[30]

World War II delayed the inevitable transition for a few years. Even before the US entered the war, the government began regulating production of materials toward what was necessary for national defense. The US War Production Board, formed in January 1942, began regulating the production of all locomotive builders. Diesel locomotive manufacturing was severely restricted because diesels used large quantities of copper and high-strength steel that were needed to build submarines and other naval vessels. This restriction benefited Alco, Baldwin, and Lima because,

---

[30] Railroads had to contend with locomotive engineers' unions, which they feared might claim those were really four separate locomotives that needed four crews. So EMD made them in two versions: "A-units" that included a cab and "B-units" that did not. Since the B-unit could only operate when connected to an A-unit, the union would be forced to acknowledge the combination as only one locomotive.

although railroads clamored for diesels, they had to accept many more steam locomotives.

The WPB did not restrict diesel locomotive research, but none of the three steam builders took full advantage of this opportunity. Their executives apparently saw wartime orders as signaling a resurgence of steam rather than being a temporary detour forced by the necessities of war.

That's not to say those companies did no diesel research. Starting in 1940, Baldwin attempted to develop a 6,000-horsepower diesel locomotive to compete with EMD's FT, but it was never successful. They designed it to use eight diesel engines of 750-horsepower each. Unlike EMD, Baldwin copied the approach they had used for steam. They tried to mount all eight engines onto a single frame. It was a classic example of how a market leader can be unable to think beyond their legacy core of expertise. Even with only four engines installed, the resulting behemoth never worked, and by 1943 they gave up on it. Their follow-on locomotive, informally known as the Centipede (so named because its nearly unbroken line of twelve axles under the carbody made it resemble the slender arthropod) was only marginally more successful. It sold a total of only fifty-two units. At least Baldwin had learned a lesson and designed the Centipede as a coupled set of two units of 3,000 horsepower each.

As late as 1946, a Baldwin executive predicted, "Better steam locomotives than we have ever known are on their way," but railroads knew otherwise. Diesel locomotives were far more economical to operate, so once the war was over and wartime restrictions rescinded, they stopped buying steam locomotives almost overnight. The last commercially manufactured steam locomotive for a US railroad was built by Lima in 1949.

After the war Baldwin and Lima were never again profitable. They merged in 1950, but that only delayed the inevitable. In 1956, the company then known as Baldwin-Lima-Hamilton produced its last diesel locomotive, closed production, and sold its assets to the meat-packer Armour & Company.

Alco had the best opportunity to compete with EMC. Even during the Depression, it had sufficient funds that it could have invested in research. Its first diesel switch engines were introduced four years before EMC's. According to Dr. Churella, in 1935 Alco held 83 percent share of the US diesel locomotive market. After EMC's introduction a year later, Alco's share fell to 26 percent and never recovered.

Alco had a well-positioned advocate for diesel technology among its leaders. Perry T. Egbert joined Alco in 1921 after receiving a degree in mechanical engineering from Cornell University and serving as a pursuit pilot for the 185th Aero Squadron in France during World War I. He became the manager in charge of Alco's diesel development in 1929 and led the company through a number of firsts in diesel locomotives.

Besides introducing the diesel switch engines that dominated the market in the early 1930s, Egbert led Alco through the introduction of the industry's first "road switcher," the RS-1, in 1940. This came about because the bulldog-nosed streamliners popular at the time gave engineers excellent views facing forward but no view of anything behind them, making them impractical for switching work. Alco took its existing end-cab switch engine, added a short hood on the other side of the cab, fitted it with crew accommodations and a toilet, and geared it for higher speed operation. The result was a locomotive useful for both switching work and over-the-road transport. It was immediately popular, but World War II delayed its widespread adoption. The RS-1 and its successors had no serious competition until 1949, when EMD introduced its first viable road switcher, the GP7.

Egbert also led Alco's launch of their competitor to EMD's E- and F-series of streamliners, the Alco FA (freight) and PA (passenger) series of streamliners, introduced in 1946. Like all other Alco diesels, these used traction motors built by GE. Until 1953 this was under a formal contract between the two companies. In 1950, Alco transferred all sales and marketing of diesels to GE. When GE management terminated the sales and marketing agreement three years later and began working on their own line of diesels, Alco had no sales or marketing staff to take it over. By the early 1960s, EMD's primary competitor was GE, not Alco.

In the 1930s and 1940s, Alco's senior management showed little support for Egbert. Dr Churella says, "It seems almost as if senior Alco executives punished Egbert for his training in diesel locomotion and his lack of commitment to steam power." One problem was that Alco did not have a chairman of the board for seven years, from 1933 to 1940.

In 1938, Alco president William Dickerman claimed, "The possibilities of the diesel-electric locomotive are already fixed and known...not so with the steam locomotive." He then went on to list a series of upcoming improvements such as roller bearings and streamlining that were no more than incremental innovations in a world being overcome by a disruptive innovation.

By the time Egbert's value was recognized and he was promoted to president of Alco in 1953, it was too late to counter the EMD juggernaut. Alco survived in the locomotive business until 1969 primarily because EMD knew it needed a competitor to avoid being declared a monopoly by the US government. Even so, the government brought antitrust charges against EMD in 1961, but they were dismissed by 1967. During testimony, legislators heard Baldwin and Alco executives admit EMD became dominant because they were still focused on steam while the world was moving to diesel.

Neither GM nor GE are in the locomotive business today. GM sold EMD in 2005, and GE sold their locomotive business in 2018. Both sales were primarily conducted to generate cash that the companies needed to cover losses in other businesses.

CONCLUSION

This classic study of disruptive change demonstrates another key tenet of this book: innovation is not monolithic. The principles and approaches that work for securing incremental innovation are distinct from those that guide high-stakes decisions amid disruptive changes to culture and industry. Today's managers, over the course of their careers, will need to understand how to foster environments conducive to both types of innovation. Now that we have seen the impact a disruptive innovation can have on an industry, it is time to learn how to manage it.

Key Takeaways for Disruptive Innovation in the Railroad Industry

1. Steam locomotive manufacturers worried that spending money developing diesels would be more likely to drain profits than increase them, so they resisted the change. EMD, not having a steam locomotive business to cannibalize, saw every sale of a diesel locomotive as a step on its path to growth. It was yet another example of the curse of the corporate business model.

2. Disruptive innovation needs to be driven by leaders who can be strong advocates for it with senior executives. EMC had those leaders in Harold Hamilton and Charles Kettering. They refused to accept a "no" from GM's CEO and pressed forward. Alco's champion, Perry Egbert, was unable to convince senior managers who were steadfast supporters of steam that diesel needed increased investment.

3. Disruptive innovation should be structured as an entity separate from the organization's core business. General Motors achieved success by keeping EMD as a separate division. Alco and Baldwin failed because they tried to drive diesel development through their core businesses.

DISCUSSION QUESTIONS

1. Compare this story of the transition of locomotives from steam to diesel to the story of today's automobile manufacturers facing the

emergence of electric vehicles. What could automobile manufacturers learn from this earlier story?

2. None of the three steam locomotive builders were successful in their attempts to build diesels in facilities designed for steam. What could they have done differently to stay relevant?

3. How could Baldwin have used their purchase of Whitcomb Locomotive Works to take a leadership role in the transition from steam to diesel? Why do you think they didn't do that? Why you think they purchased Whitcomb?

4. The 1923 Kaufman Act is an example of how government legislation can help drive a disruption. If New York City had not banned all steam locomotives from the city, what impact do you think that would have had on the timeline of the transition from steam to diesel?

5. What examples exist today where government legislation is having an impact on innovation either positively or negatively?

6. When studying a historical case such as this, it is easy to conclude that the leaders should have seen the disruption coming and mobilized to deal with it. This is not easy to do when you are in the middle of a disruption currently underway. Besides the transition to electric vehicles, what are some other examples of disruptions underway in today's markets? Are the market leaders doing enough to assure they will retain their leadership after the disruption is complete? If not, what should they be doing?

Chapter 7

Creating an Environment for Disruptive Innovation

The transition of railroad locomotives from steam to diesel is an example of a disruptive innovation that took several decades to complete. Clayton Christensen, in his classic book *The Innovator's Dilemma*, shares other examples of disruptive innovations that encroached slowly but steadily on existing markets. Large steam shovels were eventually displaced by mechanical excavators built by companies that had entered the market selling small backhoes attached to tractors. Inkjet printers initially targeted for cost-conscious users of personal computers evolved to where they could replace more expensive laser printers in many applications. In both cases the new products entered at the low end of their markets, which was not a priority for the leaders. As technologies matured, these new solutions claimed increasing shares across broader segments of their markets.[31]

---

[31] At least Hewlett-Packard hedged their bets by investing in both inkjet and laser-printing technologies. And they were smart enough to set up the inkjet printer business as a separate division in Vancouver, Washington, that competed with rather than collaborated with the laser printer division in Boise, Idaho—a situation I had to navigate delicately when I was leading both divisions through the process of deploying a standardized, company-wide version of a disruptive innovation in electronic manufacturing technology.

Not all disruptive innovations follow that slow, evolutionary path to success. Some burst onto the scene like a flash of lightning. The introduction of the DCA described in chapter 3 is an example of one that occurred quickly. Why the difference? Consider two scenarios:

- **Innovations targeted at markets already served by well-established competitors.**
  The transition of locomotives from steam to diesel is an example of this scenario. If the challenger, EMC, had started by attempting a direct frontal attack on the core business of the market leaders—mainline steam locomotives—it would not have succeeded. Leaders are well motivated to protect their core businesses and have the money to do so. In this case, diesel locomotives initially provided less value, not more, to those core customers. Even if diesels had been marginally better, most customers would have stayed with what they know. The phased approach works best here. Start by providing measurable value to a segment of the market that the leaders do not consider important and that is ripe for capture. Once you have established this foothold, grow by expanding your solution to deliver competitive advantages across broader segments of the market.

- **Innovations targeted at emerging markets with no entrenched competitors.**
  The DCA is an example of this scenario. Although the fiber-optic market was served by a different solution, it was far from ideal. Customers had no allegiance to that solution and were receptive to something better. In this case the best approach is to deliver a solid solution from the outset that clearly meets the needs of those customers better than the alternative. This is not the time to deliver only a partial solution. Doing so gives a competitor a window to introduce its own product that attacks your weaknesses. That competitive product might not be any better, but if offered by a

market leader, most customers will choose to stay with a supplier they already know.

DISRUPTIVE INNOVATION ORGANIZATION STRUCTURE AND MANAGEMENT

Counting on the leaders of an existing business to drive disruptive change—an approach that works so well for incremental innovation—is unlikely to succeed (with one notable exception we will explore in the next chapter). Disruptive innovation takes time—as long as several years to deliver results. If the same managers who are directly responsible for meeting monthly and quarterly results are tasked with managing the innovation project, they will have little allegiance to it. As an activity that does not contribute to this month's profits, they will treat it as the first to be cut if performance is lagging. To avoid this scenario, disruptive innovation needs to be carried out by a team not held hostage to monthly profitability. Here are four ways companies have done this successfully:

1. Central Research Laboratory

Many multinational corporations manage disruptive innovation through a central research laboratory that conducts fundamental research on innovative new technologies. The classic example is Bell Labs, created by Alexander Graham Bell in 1880. Originally part of the Bell System, Bell Labs has transitioned through various owners, including AT&T, Alcatel-Lucent, and Nokia Corporation, where it currently resides. Over the decades, Bell Labs researchers have contributed to such groundbreaking developments as broadcast television, stereophonic sound, transistor technology, solar cells, laser technology, and optical fiber transmission technology. As of 2024, Bell Labs researchers have received ten Nobel Prizes.

Hewlett-Packard Laboratories, whose researchers developed the high-speed photodetectors we used in the DCA program, was another example of a corporate research laboratory dedicated to developing fundamental technologies of value to company businesses. The original HP Labs split each time the company split and today consists of labs in at least four separate companies: Hewlett Packard Enterprise, HP Inc., Agilent Technologies, and Keysight Technologies.

Today, many large corporations have similar labs. Examples include typical players in high tech—Amazon Lab126, X Development LLC (formerly Google X), and Microsoft Research, for example, plus labs in such diverse markets as automotive (Volkswagen Automotive Innovation Lab) and retail (Lowe's Innovation Labs, Kohl's Innovation Center).

You will sometimes hear this type of lab called a "Skunk Works" after the Lockheed (now Lockheed Martin) lab created by the famous Kelly Johnson. Since being formed in 1939, the Skunk Works has developed such notable military aircraft as the P-38 Lightning, SR-71 Blackbird, and F-35 Lightning II. The name "Skunk Works" is a trademark of the Lockheed Martin Corporation, so to avoid trademark issues, we will not use that name here.

Some companies contract with outside innovation firms to accelerate their innovation process. Accenture Innovation Centers and Porsche Consulting Innovation Lab are two examples of innovation labs that conduct contract research for outside clients.

Corporate research labs are not always responsible for delivering fully functional products ready for sale to customers. They may conduct fundamental research on technology, then share that knowledge with separate R&D teams in the company's business units. Those business-managed R&D teams develop the commercial products or services that derive from that research. This approach, while not universal, allows corporate researchers to focus on invention rather than development.

Over most of the twentieth century, corporate research labs tended to innovate in isolation, but in recent years things have changed. Today it is financially challenging for a corporate research lab to fund disruptive innovation entirely on its own, and senior executives expect it to contribute to company profits sooner than in the past. As a result, the work it does today tends to emphasize later-stage development rather than blue-sky research. To fill this gap, the trend has been along two lines:

- Establish relationships with universities to codevelop technologies of joint interest (Kumamoto 2017).
- Help fund promising start-ups with an eye toward acquiring their technologies.

Evidence suggests this new approach has not been as productive as the old one. In a widely cited paper presented at the 2019 conference of the National Bureau of Economic Research, Ashish Arora and colleagues at Duke University reported:

The past three decades have been marked by a growing division of labor between universities focusing on research and large corporations focusing on development. Knowledge produced by universities is not often in a form that can be readily digested and turned into new goods and services. Small firms and university technology transfer offices cannot fully substitute for corporate research, which had integrated multiple disciplines at the scale required to solve significant technical problems. Therefore, whereas the division of innovative labor may have raised the volume of science by universities, it has also slowed, at least for a period of time, the transformation of that knowledge into novel products and processes (Arora et. al. 2020).

Not every disruptive technology will come from a corporate research lab—many corporations do not have such a lab, and even for those that do, budgetary limitations will prevent that lab from investigating every possible innovative idea. There is value in letting business units work on disruptive innovation projects that are important to their own businesses but lower in priority for the overall corporation.

The challenges involved in managing a corporate research lab are enough to fill an entire book by itself and are beyond the scope of this one. Refer to the references section at the back of the book for more insight.

2. Corporate Venture Capital Approach

In recent years it has become more common for companies to leave disruptive innovations to start-up companies and create a "corporate venture capital" (CVC) operation that invests in the most interesting of those start-ups (Rinker 2022). The goal is to eventually acquire the successful ones or their technologies. Figure 7.1 shows that this type of investment has grown over 30 percent per year since 2010.

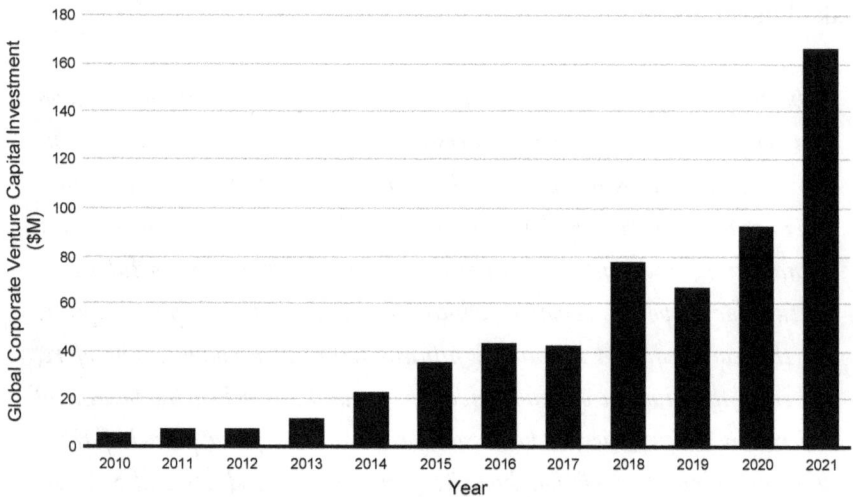

Figure 7.1: Global corporate venture capital investment by year
(adapted from Bain & Company, McKinsey & Company data)

The track record of corporate venture capital departments has been mixed. Most start-ups fail. Traditional venture capital organizations understand and accept this. They know they need to be in it for the long term—years, not months. They are confident that over time, the return on their investment in successful start-ups will more than offset losses from the failures.

This is a foreign concept to many corporate executives, who fear every failure will be a black mark on their records—a classic case of the curse of the corporate business model. Many companies that have tried the corporate venture capital approach eventually shut it down because of what they see as a dismal track record.

Many failures occur because corporate executives don't understand that a venture capital operation needs to be managed differently than a revenue-producing business unit. If the expectation is that the CVC must deliver profitable results within the first year or two or be considered a failure, it might as well be shut down immediately. That will not happen.

Here are four ways to optimize its chance for success:

- **Assure that the CVC has the full support of senior management.** This is not as easy as it might seem. Senior management must understand that this is not a short-term commitment. The CVC may not deliver positive results for several years and will undoubtedly invest in more failures than successes. The C-suite must accept this risk and commit to a stable, multiyear budget that will not be cut in the first business downturn. While regular reviews are important, their purpose should be to inform the C-suite of status rather than to let the C-suite decide whether to fund or kill projects. Leave those decisions to the CVC team, not senior management. It would be well for the C-suite to consider the CVC's total budget to be a sunk cost, with any positive returns being an unexpected benefit.
- **Structure the CVC as a separate organization.** It should not be something done part time by individuals who have other responsibilities. It needs an explicit budget and a named leader. Every member of the team must have a specific role that everyone understands.
- **Clearly define the CVC's objectives.** Is it to identify early-stage start-ups that are developing potentially disruptive technologies, or is it to identify later-stage start-ups that are ripe for acquisition? Those are different objectives. In the first, the CVC gets exposure to disruptive innovations several years before the rest of the world, with the opportunity to get locked in early. The downside is that

the investment risk is high, and well over half the investments will not pan out. In the second, the risk is lower, but the opportunity to acquire truly disruptive technology is also lower—the most innovative start-ups will already have relationships with early-stage corporate investors. The objectives should also spell out the CVC's investment priority. Is it to establish relationships with start-ups that align well with the company's existing businesses, or is it to explore opportunities in new market spaces?

- **Don't hold the CVC to the corporation's standard budgeting process and cycles.** Start-ups march to a different schedule. They are often short of cash and need a quick infusion of money to keep the lights on. When the opportunity to fund a start-up arises, the CVC needs to act quickly. Holding off for even a week waiting for executive approval can jeopardize the opportunity. Establish the CVC's approved overall budget in advance, then give its leaders wide latitude for how to invest it. If the executive staff is not comfortable placing that much trust in the CVC's leaders, either find new leaders or shut the organization down.

Research conducted by Stanford Graduate School of Business shows a wide variation in how corporate venture capital organizations are structured (Strebulaev and Wang 2022). The best of them operate similarly to traditional venture capitalists. They focus on early-stage start-ups and may talk to dozens every year. While they may only invest in one or two, by talking to all the others they get early insight into potentially disruptive market trends. The few investments they make give them access to promising mergers and acquisitions (M&A) targets a few years down the road.

I will finish the discussion of CVC programs with insight from someone who has lived on both sides of the process: my son, Greg Hinch. He has worked in start-ups seeking corporate investment and in corporations that have acquired start-ups. He is now CTO of a software company. His thoughts:

I've been around this on both sides over the years. In my opinion, corporate VC is bad for both sides and rarely works well. The start-ups think their idea is validated because they have a "customer" with the corporation, and as a result they often don't spend enough time talking to their real customers. Meanwhile, the people running the corporate venture capital operation are quite disconnected from the actual operations of the business, and the start-up does not get any traction within the corporation (let alone find other customers). Meanwhile, the corporation spends a lot of resources finding and investing in start-ups, most of whom fail, and the most likely outcome is a few hires through acquisitions (a very expensive way to recruit talent!). I think corporations that want more exposure to start-ups these days should do one or both of two things:

1. *Invest in reputable, dedicated VC funds as limited partners.*
2. *Find ways to streamline a ring-fenced procurement process which works for start-ups. A handful of people building an innovative new product aren't going to be able to satisfy an eighty-page information security document and wait for a three-to-six-month sales and procurement process (typical of enterprise software).*

3. Limited Scope Innovation Project Funded by the Business Unit

Another approach is to create a dedicated team inside the business unit whose scope is limited to developing a specific innovation for that business without any broader responsibility for innovation in other areas. That is how we organized the DCA project of chapter 3.

For this approach to succeed, the innovation project must be structured as a separate group insulated from the business unit's near-term financial results. The leader of the business unit (typically a vice president or above) must be committed to accepting this level of risk. This suggests the innovation project should be small enough that it will not have a significant impact on the business unit's overall results. Otherwise, whenever results are lagging, it would be too tempting to see this as the first place to cut. Projects that require a larger investment, assuming they have high enough potential, would need to be funded at the corporate level.

4. Corporate-Level New Business Creation Group

The above approach of having the business unit staff a limited-scope disruptive innovation project can be effective, but it is not the most efficient one for a large multinational corporation. The expertise gained by that project team doesn't easily transfer to innovation projects across the rest of the corporation. When learnings from one disruptive innovation project are applied to projects in other businesses, those projects become more efficient.

This limitation can be addressed by adopting a variation of the corporate venture capital approach. Traditionally, the CVC group funds outside start-ups and acquires the most successful ones. In this variation, which I call the corporate New Business Creation (NBC) group, it funds disruptive innovation projects inside the company.

The NBC group's objective is to foster the development of disruptive innovations that benefit the company but would not otherwise have the necessary support—not because of technology or market challenges, but because the priority of the business units must be on near-term performance. Similar to an outside venture capital fund, the NBC group controls a predefined level of investment and allocates it to the most attractive disruptive innovation ideas across the entire company.

The NBC team should focus on the early stages of the development process—from crafting the initial concept through an "incubator project" phase that answers fundamental questions of technology and market opportunity. From the outset, the next steps to be taken if the incubator project is successful must be clearly understood. There are typically two alternatives. If the project will align with an existing business unit, the product line that will benefit from it should take it over and carry the project to completion. If it does not align with an existing business unit, the corporation must be committed to creating a new business unit for it. In rare cases it might make sense to license the technology or sell the product to another company, but these alternatives are not as valuable and should be used sparingly. If neither approach has a strong advocate inside the company, the project should be terminated before major expenditures are incurred.

An example of how this can work is shown in Figure 7.2. The NBC group drives the first four phases of a five-phase process. The first step is to collect ideas for disruptive innovations. These can come from internal businesses, from sales organizations, from the corporate research lab, or from outside sources such as university researchers or venture capitalists.

In the second step, the NBC group constructs the list of possible projects. The project descriptions may not include significant detail—not even much of a vision of a final product. At this point the objective is to understand the market opportunity and the technical challenges, so the project descriptions need enough information to assess their relative priorities. This should include a description of the target market, the unmet customer needs the project would address, the most important questions to be answered, and an estimate of the staffing, budget, and time needed to get those initial questions answered. If the project would eventually fall under the umbrella of an existing business unit, this should also be indicated.

The third step is to rank those projects using a priority-setting process that includes representatives from the NBC group, the corporate research lab, and the corporation's existing businesses. Decisions should be based on a combination of expected return on investment (ROI) and likelihood of success. Some of the funded projects should support existing businesses, but some should explore blue-sky ideas that do not obviously align with any existing businesses. In all cases the funded projects should be disruptive innovations that would not otherwise have been funded. The NBC needs to avoid being no more than a source of additional funding for incremental innovations within the business units.

The fourth step is for the NBC group to staff a small "incubator phase" project for each funded priority. The objective is to validate questions of technology and market opportunity before incurring large expenditures. Depending on the result of this phase, a successful incubator project will

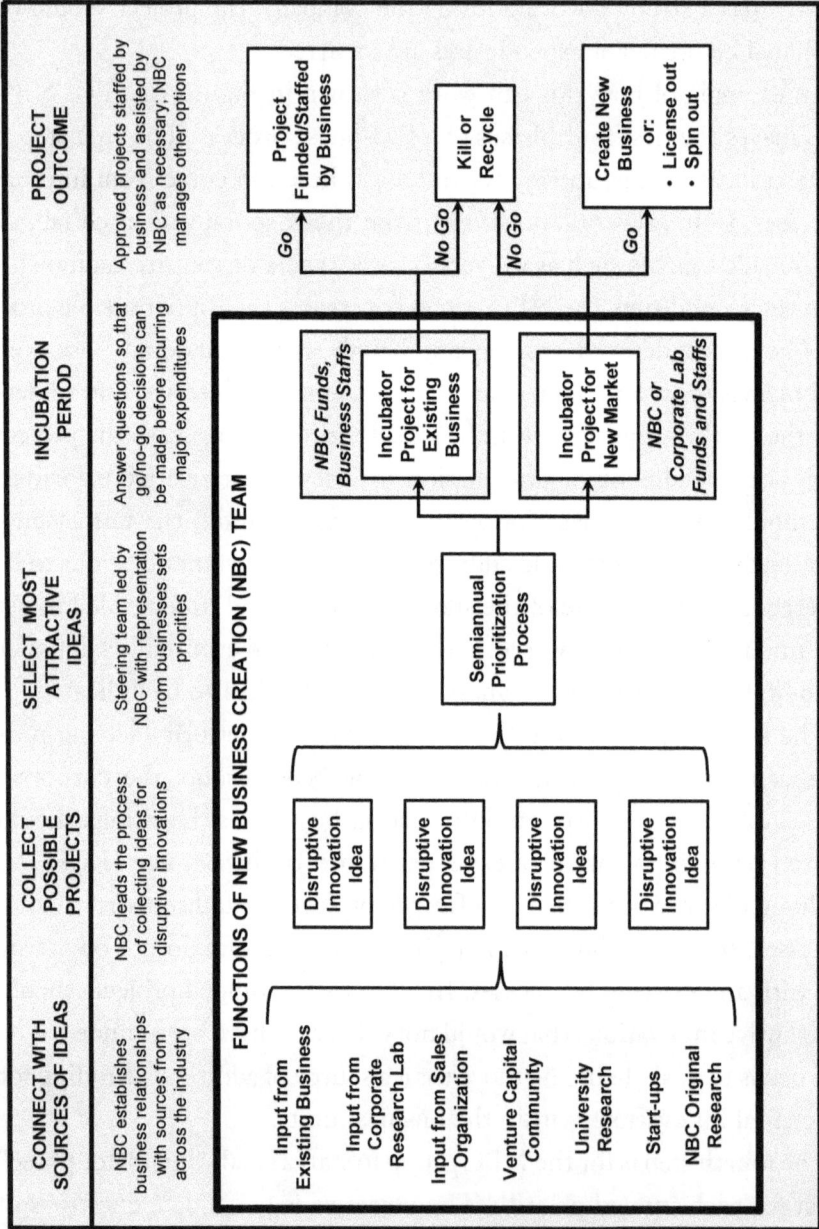

Figure 7.2: Example of New Business Creation operation

either be picked up by an R&D team within an existing business unit or by a newly formed business unit.

Not every project will make it successfully through the incubation phase. Eighty percent of start-ups fail. If you are diligent about following the process, your results may be better, but you should still expect half or less to pass this phase. That is the whole reason you have this phase—to weed out the failures before making large investments. The NBC team should not be measured on the number of failures it funds, but rather on the number of successes it launches and its overall ROI to the company.

The corporation's C-suite needs to set the NBC group's overall budget and expected ROI. Since representatives from existing businesses will strongly advocate for projects that support their own businesses, the NBC group needs to decide in advance what portion of the total budget will be assigned to new-market blue-sky ideas. The result of the priority-setting process will be a set of projects whose total funding fits within the budget.

The NBC group needs to work closely with the corporate research lab and the business units to (1) facilitate deployment of technologies developed by the corporate lab into products for the business units and (2) share with researchers in the corporate lab new technologies the business units want to see investigated. The NBC group should not set priorities for the corporate research lab. While it will provide inputs, lab researchers should be free to explore true blue-sky innovative ideas that may not have obvious benefits to any current business. If these researchers only work on projects requested by the businesses, they will be little more than an extension of business R&D teams—more focused on development of commercial products than on true innovative research.

The NBC group should have the ability to explore disruptive innovations in entirely new markets and also across all the corporation's current businesses. This can be a source of friction if those businesses feel left out of the innovation process. To avoid this, the NBC group needs to work closely with those businesses to identify opportunities of interest that are outside the business unit's ability to staff and fund.

Although rare today, the NBC approach is being used successfully. As one example, IBM has set up an incubator structure they call Area 631. It funds short-term, small-group innovation projects (six people working for three months on one innovative breakthrough) in a format similar to the incubation phase of the NBC operation. IBM launched the first Area 631 project in Canada in 2018. Its success has led IBM to establish eight more Area 631 programs across the company (Astorino 2020). As another example, Geoffrey Moore, in his book *Zone to Win*, describes a similar approach Cisco Systems has used to fund disruptive innovation internally through their Emerging Technologies and Innovation group.

DISRUPTIVE INNOVATION RULES FOR SUCCESS

Regardless of how a disruptive innovation project is structured, it should be managed according to the following rules for success:

- **Staff the project with a restricted number of talented resources.** One of Kelly Johnson's rules for the Skunk Works was to restrict the number of people on a project: "Use a small number of good people." At the beginning, a disruptive innovation project should have a very small staff—only the few key people who can answer fundamental questions about the project's feasibility. Once those questions have been answered, the project can add staff as necessary in R&D, software development, marketing, and for physical products, manufacturing development. These should all be dedicated resources. If any of them also retain responsibility for delivering monthly results, they will inevitably end up focusing on the near term rather than on innovation.
- **Designate one person as the project owner.** This person needs to be someone with the skill and confidence to advocate for the project with upper management and not be afraid to push back when the team is not getting the necessary support. He or she needs to be a good storyteller both up and down the organization—able

to sell the project to upper management and inspire the staff who are working on it. Everyone on the project needs to know who owns this role and be prepared to accept their decisions.

- **A senior executive (vice president or above) must sponsor the project.** While this executive will undoubtedly also manage company businesses that must meet monthly profit targets, he or she must be fully committed to keeping the innovation project insulated from those monthly metrics.

- **Staff the project only with people who are fully committed to it.** Nothing will cause a project to fail faster than staffing it with people who are only there because they were told to be rather than because they want to be.

- **Manage the project to explicit metrics.** This involves regular tracking of two types of metrics: financial and technical. On the financial side, the project team should understand they have not been given a blank check for unlimited spending. The leader should negotiate a budget with the executive sponsor, then manage the project to that number. Initially, the agreed budget may only cover the next quarter or two—just long enough to assess whether the project is likely to be successful. As the project moves forward and the possibility of success improves, a longer-term budget can be established.

    On the technical side, the project should be measured by tracking how well it delivers results against predetermined, explicit milestones. In the early phases, this may be as simple as getting answers to fundamental questions of technology and market opportunity. During this initial period, the milestones being tracked will be no more than a few weeks away. If results look encouraging, the next round of metrics should be more detailed and extend over a longer period of time.

    These metrics need to be reviewed regularly—in formal meetings no less often than monthly. If the project is not meeting either its financial or its technical goals, take immediate action to understand why. This does not mean killing the project or sacking the leaders. Disruptive innovation is rarely a predictable process. This is especially true in the early phases. As the team begins getting results, there will almost certainly be changes in direction. These should not be considered failures but rather as being part of the learning process. If the results do not look encouraging over an extended period, there may come a time to make the difficult decision to pull the plug on the project.

- **Succeed fast; fail faster.** Most truly innovative ideas do not pan out. If your failure rate is low, you are not being aggressive enough in pursuing novel ideas. This could be because people are afraid of the punishment that may come with failure, afraid they would not get the necessary funding, or possibly because they do not see value even if they are successful. It is the responsibility of the organization's senior leadership team to remove those obstacles.

Let's take a closer look at this last point. It has particular significance for our efforts to understand how to prime a work environment for disruptive innovation. Given the expected high failure rate of disruptive innovations, you must determine an innovation's fate quickly—before you have invested too much time and money into something that goes nowhere. Conversely, if the innovation proves successful, you will want to start reaping benefits from it as soon as possible. Here are four things that will help you achieve both objectives:

1. Map out the key questions to be answered and the difficulty of answering them. Then answer the most important questions first.
2. Use the concept of minimum viable product (MVP) to enter the market with an initial solution quickly.
3. Consider using Agile methodology for software projects.
4. Consider using platforms to accelerate development.

Let's explore each of these points now:

Map Out the Key Questions to Be Answered and the Difficulty of Answering Them, Then Answer the Most Important Questions First

This might seem illogical; it would seem smarter to do easier things first so you can build your confidence and show tangible results to your managers. But think about it. Why waste time creating a flashy user interface or a sleek industrial design when you don't even know if the product will work or whether there is really a market for it? Get the most important questions answered first.

Use the Concept of MVP to Enter the Market with a First Solution Quickly

Your first solution may not satisfy many customers, but it should be good enough to at least get the attention of some. Consider using the concept of MVP to get that solution to market quickly. The MVP concept gets a lot of flak these days, but it can be a powerful tool if you understand how to use it properly. The problems I have seen arise because the "MVP" as defined did not even qualify as a "minimum" viable product.

The basic premise of MVP is that the first product you introduce should have enough features to attract an initial set of early-adopter customers but not everything you think it should eventually have. These initial customers will help you validate the idea and give you feedback on how to improve it. Some authors contend the only objective of MVP should be to get that feedback from customers. I have a broader view that recognizes the ability of an MVP product to quickly generate significant revenue from specific market segments.

The Odin project of chapter 3 is an example of an MVP designed to do little more than get feedback from customers. It was a quick-turnaround software project that gave us the opportunity to talk to customers and learn what they really wanted. We never expected it to be profitable by itself, but we saw it as critical to assuring the eventual success of BudLight.

BudLight was also a minimum viable product. Its first release included only two modules and served the needs of only one segment of the fiber-optic market. As we continued to introduce new modules for other segments, the product grew to become the unquestioned leader across a much broader market.

MVP is not a new concept. Consider, for example, the 1934 *Burlington Zephyr* trainset's record-breaking run from Denver to Chicago described in chapter 6. It was the result of an MVP locomotive design. A design engineer had to ride in the cab the entire way to deal with such challenges as a broken traction motor bearing, burning insulation, a nonfunctional starter, and stuck brakes—all typical of a design rushed into production with unresolved problems. But the result was unambiguous. That one run convinced railroads of the value of diesel locomotives, and they never looked back.

The MVP approach won't be successful if you aren't committed to solid follow-on plans. That first product will rarely be exactly what customers want. If you don't follow through with improvements based on the feedback you get from early adopters, you will alienate those customers who took a chance on you.

Consider Using Agile Methodology for Software Projects

The concept of MVP applies equally to software projects, but software has additional options not readily available to hardware projects. With hardware, each iteration of the product can take many months to design, assemble, and test. With software, updates can be released more often, which allows new capabilities to be added regularly. This can be a powerful way to get an initial product to market quickly.

The software development approach known as Agile methodology is ideal for doing this. Agile has several variations, but a common thread is that new features are released through regularly scheduled software updates called "sprints." Rather than defining a lengthy list of features and delaying the release until all those features have been developed, Agile defines "sprints" with specific release dates (often either monthly or quarterly). Each sprint launches only the features that are ready at the time of the release. Features not ready for the current release are moved to future sprints.

Agile methodology allows software developers to more easily interact with customers to learn which new features are most important. These can be added to the software development road map, which can be shared with customers to show where those features fit on the road map. Customers have confidence that even if their most desired feature is not yet available, they know which release should have it. Agile is also the best approach to use when the desired end result is uncertain and you need flexibility to experiment with different alternatives.

Agile is not perfect. Because it focuses on speed (it is no coincidence that feature releases are called "sprints"), planning work and documentation are often minimal. If a feature is taking too long to develop, the typical response is "No problem, we will move it to the next sprint."

That is not to say planning is unimportant in Agile. It is still very important but focuses primarily on the next sprint rather than the entire project. This makes it hard to predict which features will be available at any given time. And Agile depends on getting feedback from customers to learn what they consider important and what they think of the capabilities already developed. If customers are not willing or able to do that, the feedback loop will be broken, and the software team will be flying blind.

My son, Greg, is CTO of a software company. He has extensive experience managing Agile software projects. Here are his thoughts:

Sprints can be monthly or quarterly, but I've always worked in even shorter iterations (two weeks is typical). There has been some pushback on this short cycle in recent years, but it's still common. It's important to find ways to ship work frequently and not keep carrying it over into the next sprint. There's no point in adopting an Agile methodology if work gets stuck at the end of the process before it gets to customers.

One of the biggest challenges I see large organizations struggle with when adopting Agile is they butt up against (mostly artificial) deadlines. Agile is iterative and discovery-led when successful, and if senior management is used to setting deadlines to feel like they're motivating people to work hard, you lose most of the benefits (and make the team pretty grumpy along the way!).

Agile is not the best approach for every project. The alternative is the traditional method known as Waterfall. Waterfall was originally used to manage hardware projects and is still very important in that venue. In Waterfall, the full list of necessary features is defined at the beginning of the project. The project then moves sequentially from one phase to the next at the pace necessary to keep that full feature set intact. The transitions are controlled by a series of checkpoint reviews with key stakeholders to analyze the project's status versus goals and confirm it is ready to proceed to the next phase. You will sometimes see the checkpoint reviews called "stage gates," a term trademarked by Stage-Gate Inc. (www.stage-gate.com).

Waterfall relies on comprehensive planning to lock down the project scope up front and clearly delineate exactly what deliverables are to be included. Once this is done, making changes, while not impossible, can be difficult.

Waterfall methodology is necessary in environments where the final product is in a heavily regulated industry or where the product needs all the features from the outset—you wouldn't want to fly in an airliner where the software development team said, "It is taking longer than expected to develop the navigation software, so we will leave that out of our first sprint."

In many cases a hybrid approach works best. A core set of features is defined that must be included in the first release, and the project will take as long as necessary to develop those. This first release may use a modified Waterfall approach that includes a detailed planning step to define the necessary minimum set of features. It can use Agile sprints to deliver individual features, but only after the necessary minimum set is complete is the code released. Once that initial release is done, additional features can be introduced through a series of Agile sprints.

Consider Using Platforms to Accelerate Development

In their book *The Power of Product Platforms*, Marc Meyer and Alvin Lehnerd define a product platform as:

> *A set of subsystems and interfaces that form a common structure from which a stream of derivative products can be efficiently developed and produced.*

The value of a product platform is that by serving as the foundation for multiple products, it can accelerate design time and reduce the number of people needed to develop each new product based on that platform.

Platforms can have multiple tiers of standardization (see Figure 7.3 for an example I have used in the electronics industry). At the lowest level are the common building blocks such as electrical or mechanical components and software utilities such as printer drivers. An intermediate tier may consist of integrated subsystems such as fully functional flat-panel displays, power supplies, or functional software modules. The top level consists of product-level platforms that serve a variety of needs. As you move up the levels, leverage increases and efficiency improves. This gives you the ability to release products more rapidly because you have less original design work to perform. The trade-off is that as the level of standardization increases, the flexibility to make changes from the standard decreases.

Platform Advantages

- Once a platform has been created, new products can be introduced more rapidly and with less development cost than a custom design.

Hosted Instrument	Midrange Instrument	High-Performance Instrument	Low-Cost Instrument	Specialty Instrument

PRODUCT PLATFORMS

Increasing Leverage
Decreasing Flexibility

Electronic Subsystems	Mechanical Subsystems	Front Panel Subsystems	Power Supply Modules

COMMON SUBSYSTEMS

Increasing Leverage
Decreasing Flexibility

Measurement Software App	Calibration Software App	Electromechanical Components	Host Processing Operating System	Software Utilities

COMMON COMPONENTS

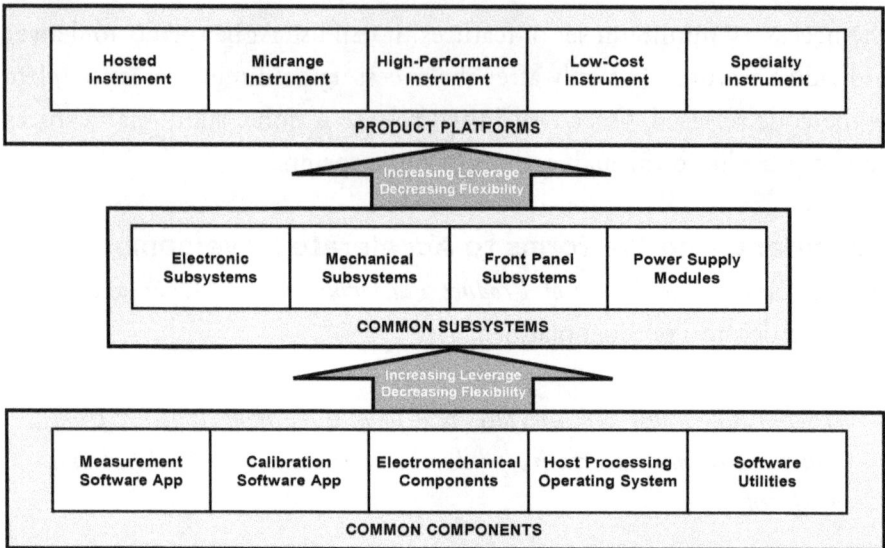

Figure 7.3: Example of three-tiered platform standards
used by a manufacturer of electronic equipment

- By reducing the time necessary to introduce each new product, introductions can occur more often, improving R&D efficiency.
- By taking advantage of predesigned component building blocks, new products can be introduced using fewer people than would be required for custom designs.
- A platform gives products a common look and feel. This is especially important when a user is likely to use multiple products based on that platform. Battery-powered tools for woodworkers are a good example. Whether you are using a rotary saw, a drill, or a sander, if they come from the same manufacturer, they will have a similar look and feel, and they will all use the same rechargeable batteries.
- When a problem arises, fixing it one time fixes it for all products that use the same platform.

Automobile manufacturing is an example of an industry that has made extensive use of platforms since the days of Henry Ford. The auto industry defines a platform as a common set of structural underpinnings of a vehicle family. The vehicle's upper body styling portion is called a "top hat." By using a common set of structural components (the platform) and different top hats, vehicles ranging from a sedan to a hatchback to a compact SUV can be designed around the same platform.

The Toyota New Global Architecture (TNGA) is an example of a modern platform underpinning a range of vehicles. The 4Runner, Land Cruiser, Tacoma, Tundra, Sequoia, and Lexus GX are all designed around a single platform. Besides accelerating the design turnaround time, using one platform has made it easier for Toyota to assure all these vehicles meet global Corporate Average Fuel Economy requirements (known as CAFE in the United States).

This does not mean the underpinnings are identical in all those vehicles. Toyota's platform has the flexibility to meet a range of needs. The 4Runner, for example, does not need as strong a frame as a full-size Tundra pickup. The TNGA platform allows for the common frame to have different lengths, material thicknesses, wheelbases, and suspensions. It is robust enough to accommodate a range of variables for the different vehicles while still contributing to a simplified design process.

Platforms are the norm for software projects. Consider the iPhone or the Android phone. In both cases, the phone is the platform, similar to how the mainframe is the platform of the DCA. Just as the DCA can be turned into different solutions by installing different plug-in modules, the iPhone or Android phone can be turned into a multitude of solutions by loading different apps.

Platform Disadvantages
While platforms offer many benefits, they also have disadvantages. Before deciding whether to create a platform, consider the following:

- A platform for physical products may not result in the lowest manufacturing cost because the platform must be designed to accommodate multiple offerings. The mainframe/plug-in module design of the DCA in chapter 3 is an example of a product whose manufacturing cost was higher than if it had been designed as a single fully integrated product. This trade-off was acceptable because it was more than offset by the reduced cost of design, the faster speed of introduction, and the improved ability to introduce follow-on products.

- A design problem in the platform will impact every product in the family, potentially dramatically increasing the cost of solving the problem.

- The common look and feel of platform-based products may not be ideal when products are targeted for different price points. The Toyota 4Runner and Toyota Land Cruiser, for example, are based on the same TNGA platform, but the Land Cruiser is significantly more expensive than the 4Runner. Standard features are not the same, so the comparison isn't exact. The perception, though, is that the two vehicles have many similarities. Even a Toyota representative acknowledged that because of this, the Land Cruiser would primarily sell to customers who were already loyal to that model.

CONCLUSION

At this point we've explored a wide range of environmental and cultural factors that impact innovation of various kinds. We've seen that while incremental innovation is based on employee knowledge of a company's values and goals, a sense of collaboration, and regular rewarding of everyday successes and productive failures, disruptive innovation requires structural change or expansion at a fundamental level. These two major types of innovation, despite their meaningful distinctions, share a premise: both benefit from the optimized, simplifying processes of high-contrast management.

Key Takeaways for Creating an Environment for Disruptive Innovation

1. A disruptive innovation can follow either of two paths. If targeted at markets already served by entrenched competitors, it should start by targeting a small segment of the market the leaders do not care about and expand from there. If targeted at emerging markets with no established competitors, it should deliver a complete solution from the outset.

2. Disruptive innovation should be carried out by a team not held hostage to monthly profitability. Four ways to do this are (1) central research laboratory, (2) corporate venture capital department, (3) limited-scope internally funded project, and (4) corporate-level NBC group.

3. The six rules for disruptive innovation success: (1) staff the project with a restricted number of talented resources, (2) designate one person as the project owner, (3) a senior executive must sponsor the project, (4) staff the project only with people who are fully committed to it, (5) manage the project to explicit metrics, (6) succeed fast; fail faster.

4. Three ways to make the development process more efficient are to (1) follow the concept of MVP, (2) use Agile methodology for software development, and (3) use platforms to accelerate the introduction of follow-on products.

DISCUSSION QUESTIONS

1. Disruptive innovations should ordinarily be targeted at a part of the market the industry leaders do not consider important. Why is this not the best strategy for markets in which none of the competitors have established a true position of leadership?

2. Some companies have established central research laboratories to lead the development of disruptive innovations. Discuss two advantages and two disadvantages of this approach.

3. The growth rate of the corporate venture capital approach of investing in start-ups has been over 30 percent per year in recent years. What is the justification for investing in a start-up rather than staffing the development internally? What are the risks?

4. What advantages does a corporate-level NBC group have over a limited-scope innovation project funded by the business unit? When would the limited scope innovation project be the better approach?

5. What are some advantages and disadvantages of building a disruptive innovation around a common platform?

Chapter 8

Case Study: Special Case of Disruptive Innovation

As we have learned, disruptive innovation should normally be conducted by an organization not held hostage to quarterly results. But there is one time when it is best conducted by the revenue-producing business unit: when results are so bad it is in imminent danger of being shut down and all the employees laid off. Often, those employees have already given up and accepted their fate, but if not, there is no one more motivated to turn things around.

That was the situation that arose in the DCA business after I had moved on to another position in the company. The case study in chapter 3 ended with the product line growing to become the unquestioned leader in its market, generating well over a hundred million dollars of revenue every year. It was growing so fast, driven by the exponential growth of the internet, that the sales manager calculated that if it continued that growth for only a few more years, every man, woman, and child in the country would own a $50,000 DCA.

As exciting as that sounded, it was clearly unrealistic. It was an example of a market that was "inside the tornado," a phenomenon described by author Geoffrey Moore in his classic book of the same name. A market's "tornado" phase is one of explosive growth as it transitions from serving a few early adopters to the first wave of mainstream users. While tornado markets sometimes move smoothly to a more stable "Main Street" phase, a more likely outcome—especially in the overhyped world of high tech—is a disastrous collapse.[32]

Sure enough, eventually the internet dot-com boom ended, causing the fiber-optic market to collapse. Not only did customers stop placing new orders for DCAs, but they also began canceling orders already on the books. As a result, in some months, net orders were zero or even negative.

The dot-com market collapse impacted businesses throughout the technology industry. After a few months of poor results company-wide, Agilent president Bill Sullivan realized the only solution was to do a massive restructuring that would include significant layoffs and close several product lines. For the second time in its history, the DCA business was perilously close to being shut down.

The DCA R&D and marketing teams understood the gravity of the situation and were not ready to give up. They knew the fiber-optic market was morbid, which meant incremental innovation in that market would never be enough. Disruptive innovation into a new market was the answer, and they had an idea.

The seeds of the idea had been sown when the marketing department discovered that the small number of DCAs still being sold were not going to customers in the fiber-optic market. They were going to design engineers in a market we had not previously considered: high-speed digital networks inside the latest generation of computers and servers. One of the most important measurements those designers needed to make was a parameter known as jitter.

To understand jitter, consider this analogy. Imagine two singers. The first is a world-class vocalist (think Loren Allred) who can hold a high note perfectly on key for many seconds. The second is a less accomplished

---

[32] The explosive growth of artificial intelligence is an example of a current market inside the tornado whose long-term future is yet to be determined.

singer who warbles annoyingly off-key when trying to hold that same note. In the parlance of digital networks, the first singer would have low jitter and the second would have high jitter. In digital networks, high jitter is not just annoying; it can lead to serious errors in transmission—you would not want your bank to record your deposit as $1.00 instead of $100.00 because jitter moved the decimal point two places to the left.

At the time, off-the-shelf commercial products were unable to measure jitter on high-speed networks very accurately, so engineers were buying oscilloscopes and inventing the necessary test setups themselves. These setups were so complex it often took hours to make a single measurement, and even then, the accuracy was questionable.

The DCA R&D team had an innovative idea for a better solution. By combining the DCA's high-accuracy hardware with new software, they were confident they could make jitter measurements accurately and quickly. The most immediate question was whether senior management would allow them to stay in business long enough to get a product onto the market. The second question was whether they could convince the industry they had the best solution.

As a result of the restructuring, I was asked to return and serve as general manager of the pared-down DCA business. The first thing I needed to do (after managing the stressful process of laying off half the 150-person staff) was to take inventory of the situation. What I found was a business with orders running only a tiny fraction of what they had been at their peak and a team whose morale was low, but one that had not completely given up. R&D was inventing the new product (which they called the "DCA-J" to highlight its ability to measure jitter), and marketing was establishing relationships with a whole new class of customers. As the team's new leader, I set three priorities for myself: (1) rebuild the team's morale so they would believe success was possible, (2) coach them as necessary to get the work done, and (3) serve as a buffer between the team and senior management to buy them the time and money they needed to complete the job.

Rebuilding the team's morale was my first step. Having just gone through a round of layoffs that cut the team in half, morale everywhere was perilously low. Everyone knew that even with all those layoffs, we were still not close to breakeven and might not get there for a year. They were legitimately concerned as to whether our business had a future. Some were already looking for jobs elsewhere. Although I could not allay all their fears, I needed to at least give them a vision of a path to success.

I pulled everyone together in a kickoff meeting for our new, downsized organization. Rather than running the meeting in my usual business-focused, engineering-style manner, I knew I needed to take a different approach—one that was uncomfortable for me because of my analytic nature. I needed the team to believe success was possible even in a situation that seemed bleak, so I decided to start with a true story from my own past—a story that might initially seem to have nothing to do with our situation, but one that had an important message for us.

Some of the team knew that in my free time, I occasionally headed out on solo adventures into remote regions of the country. The story I shared was of a trek I had done one summer morning to the top of California's White Mountain. At an elevation of over fourteen thousand feet, it was the third-highest peak in the state but, unlike similar mountains in the distant Sierras, was rarely visited. A rough dirt road wound all the way to its top, so it was not a technical climb. My challenge was that while I usually hiked near sea level, this entire fourteen-mile round trip would be at an elevation above twelve thousand feet. Afternoon thunderstorms were regular occurrences during the summer months, so I would need to get an early start and be down from the mountain before noon. I spent a day acclimating to the altitude but did not know whether that would be enough.

There was no one else at the trailhead that morning, so I started out at dawn with a few snacks and plenty of water. As the miles passed and I continued my ascent, I was surprised at how easily I was making progress.

My only concern was what might happen with the few small clouds I saw scattered about near the top of the mountain.

By midmorning, the cloud bank had grown to where it extended over my head. I was within a mile of the peak when a sudden flash and a deafening roar from the cloud directly in front of me told me it was time to turn around. But when I tried, I found I could barely move. In the thin air, just breathing had suddenly become a challenge. My feet felt as if they were encased in concrete blocks, and even the act of staying upright was difficult. I realized that with no warning, I had been overcome by a potentially lethal case of altitude sickness. I had no supplementary oxygen, so my only hope was to get to a lower elevation quickly. But that seemed an impossible task.

Soon, lightning was flashing and thunder roaring all around me. Then a sudden hailstorm followed by a downpour of rain soaked me to the core. I was exhausted. I just wanted to lay down and go to sleep—only for a little while, just long enough to regain my strength. But I knew at that altitude, if I did, I would never get up. My situation seemed hopeless. The six miles to my car was an impossible objective. I saw no way to do it. Death was looming, but I was not ready to give up.

When faced with an impossible objective, it is time to change the objective. I scanned ahead and spotted a boulder at the edge of the trail no more than twenty feet away. I decided that as of that moment, my only goal in life was to get to that boulder. Nothing else mattered. Forget about the six-mile hike. Forget about my car. Forget about my work. Forget about my family. Forget about anything else in life. Just get to that boulder.

I cleared my mind, gathered all my energy, and started to walk. I could barely drag my feet. I had to stop to catch my breath every few steps, but as the target got closer, I knew I had picked an achievable goal. Eventually, I got to that boulder. I stopped just long enough to catch my breath, then scanned the trail ahead to pick out a new goal—another boulder just a little more distant than the first. Now, my only purpose in life was to get to that next boulder.

For five hours I repeated that process from one boulder to the next, thinking each time of nothing but that immediate goal until I finally made it to my car. As I collapsed in the driver's seat, I knew that if the only goal I could have imagined was the six-mile quest to my car, I could never have summoned the energy to do it.

My message to the team was simple. Forget about our business needing to reach a break-even point a year from now; forget about other goals that may be months away. Just pick a goal you can meet within the next month, then put all your energy into meeting it. Worry about next month's goals next month.

There was more to the meeting after I finished that story—I shared my vision for our future, what everyone could do to help achieve it, and what success would look like. But it was my story of "the next boulder" that most resonated with everyone. For months afterward, discussions of what should be the next boulder were regular topics in team meetings.

One of those boulders was to convince the industry that the DCA-J was the gold standard for jitter measurement. Again, our R&D and marketing experts put their heads together and came up with a way to do it: bring together all the different tools for measuring jitter, set them up in our lab, and invite customers to a group session where they could bring their devices and compare measurements. To assure impartiality, we also invited an expert from the US Government's National Institute of Standards and Technology to oversee the process. We called the event a "Jitterfest."

Jitter was such an important measurement that customers were eager to come. Some even brought their own test setups to compare with all the other methods. Over the course of the weeklong event, customers made dozens of measurements using six different systems. The results were unambiguous. The DCA-J not only delivered the most accurate results, but it was also by far the fastest. One attendee proclaimed jitter measurements that previously took five hours now took only five minutes. Marketing quickly invented a new tagline: "The fastest way to the right answer." They shared the story of Jitterfest in trade magazines and application notes distributed throughout the industry.

We knew we had a solution that would take the market by storm, but we were still months away from launching a commercial product. Now my job was to convince senior management to keep us funded long enough to get the product onto the market. I spent the next six months doing that: putting together forecasts for orders, revenue, and profitability that promised we would reach breakeven by a certain month, then updating them regularly as new sales reports came in. Whether or not I believed those numbers was immaterial. My promises and our history of success bought us the time we needed to launch the product.

The DCA-J excited customers to such an extent that sales vastly exceeded our projections. For years afterward the company's senior leaders heralded the DCA-J as a shining example of how to do innovation.

SUMMARY: ELEMENTS OF SUCCESS

This project was successful because of the commitment from a team that understood the consequences of failure and took the initiative to find a solution. A team of outsiders from a separate disruptive innovation group whose jobs were not in jeopardy would never have been as motivated to succeed. They would most likely have moved on to an easier challenge.

Finally, this case study illustrates the role a midlevel manager should play in helping with that success. It is not to be the technical or marketing leader of the team. Leave those roles to the first-level managers and their direct reports. The midlevel manager's role is threefold: (1) give the team confidence in their future, (2) serve as coach to assure they have a path to success, and (3) run interference to insulate them from the demands of senior management to the extent possible.

Key Takeaways for Special Case of Disruptive Innovation

1. Although disruptive innovation should normally be performed by a team that is protected from the impact of near-term results, one exception is when results are so bad the business is in danger of being shut down. If employees have not already given up, they will be especially motivated to turn things around.
2. For a business in decline, the team leader has three important responsibilities: (1) give the team confidence in their future, (2) serve as coach to assure they have a path to success, and (3) run interference to insulate them from the demands of senior management to the extent possible.

DISCUSSION QUESTIONS

1. When a business is performing poorly because its market is in decline, what actions can a manager take to inspire a team to turn things around?
2. How should the manager of a declining business deal with senior management?
3. What should a manager do if they believe the business in decline has no possible way to recover? What if they believe it does have a chance to recover?

Part III

Essential Dynamics of Innovation

Chapter 9

Market-Driven Innovation

U p to this point, we have been investigating the processes at play in incremental and disruptive innovation that will equip you to deploy them in your work environment. Now that we have a sense of the essential mechanics of innovation, let's move on to some of the specific factors and dynamics that will allow you to take your execution of innovation to a new level. We will start that process with a story to illustrate the importance of understanding your target market.

In 1921, Ford Motor Company was the largest automobile manufacturer in the United States. The Ford Model T, introduced in 1908, accounted for two-thirds of all automobiles sold in the country. But by 1927, Ford's share had been cut in half. General Motors emerged as the new leader, soon to become the world's largest automobile manufacturer.

What happened? Simply put, GM listened to customers and Ford did not. Henry Ford was obsessed with producing a single family of automobiles in large volumes on highly efficient assembly lines. He was far less interested in the desires of customers—recall the famous line from his autobiography, "Any customer can have a car painted any color he wants so long as it is black." While Ford continued producing low-cost, open-air vehicles that changed in only minor ways for twenty years, GM began offering a variety of styles for a range of customers. President Alfred P. Sloan promised buyers "a car for every purse and purpose." To fulfill that promise, GM sold cars with enclosed bodies to protect riders from the elements, gave buyers the ability to trade in used cars and finance new ones, and offered a variety of models ranging from the $525 Chevrolet to the $4,485 Cadillac, each updated every year—tactics Henry Ford believed would drive GM to ruin. Only when the results became obvious did Ford obsolete the Model T and invest $100 million to replace it with the Model A (Tedlow 1988). GM did not relinquish its title as the world's largest car company until 2008 when, struggling through bankruptcy and a government bailout (a separate story we won't cover here), it lost that status to Toyota.

Figure 9.1: General Motors 1925 advertisement (author's collection)

Ford suffered from a malady typical of market-leading industry behemoths: a belief that it knew the market better than customers did and saw little need to solicit outside inputs. That strategy can work for a while, but it opens the door for more nimble competitors to serve unmet needs.

THE IMPORTANCE OF LISTENING TO CUSTOMERS

Is listening to customers important to innovation? It is a controversial topic. Some people refer to Henry Ford's alleged quote, "If I had asked people what they wanted, they would have said a faster horse." Others point to Steve Jobs's famous line, "A lot of times, people don't know what they want until you show it to them." The claim is that these quotes prove neither Ford nor Jobs felt it was important to listen to customers. Proponents of this view say that innovation—specifically disruptive innovation—should be done in isolation, away from the distraction of pesky customers. But those proponents are mostly wrong.

Let's get a couple of things straight right up front. Henry Ford never uttered that quote about faster horses, although it does appear to reflect his opinion of customers. And Steve Jobs certainly did listen to customers; he just did not depend on them to tell him what to invent.

What's the story? The problem arises when you don't understand how to do market research or what to expect from it. It won't tell you what disruptive new products to introduce. Customers can tell you their challenges, and they will often have useful ideas for incremental innovations, but rarely for disruptive innovations. Coming up with those innovations is your job.

Proponents of innovating in isolation imagine that if you talk to customers, you will just invent what they tell you to, which will be incremental, not disruptive. But listening to those customers is an essential step in your entire innovation process. By understanding their problems, you gain information that will build your creative insight.

Assume for a moment that you are doing market research for Henry Ford and have heard customers say they want faster horses. Rather than using that as an excuse to breed more racehorses, you should use it as a starting point to understand customers' true problems. Spend more time probing and you might hear something like "The speed of the horse I have is fine when it is moving, but I have to stop to let it rest every few miles. A horse needs a lot of care. I have to give it food and water every day even if I don't need to go anywhere. And by the way, I hate having to tromp through horse droppings whenever I cross a street in town."

With that perspective you will understand that breeding faster horses is not the answer. The insight you have gained will help you innovate in ways that could genuinely excite the customer. You might think of inventing a mode of transportation with a speed similar to a horse but without the need to stop every few miles, one that doesn't need daily maintenance, and one that doesn't foul city streets with solid waste. You might even imagine adding value the customer did not imagine: a mode of transportation that safely travels faster than a horse while meeting those other goals. You might think of several possible ways to do this. Innovation is the process of sorting the good ideas from the bad ones.

Not every disruptive innovation needs to come from an idea no one has ever before imagined. In your recommendations to Henry Ford, you might propose a motorized carriage powered by a gasoline engine, by an electric motor, or even by steam power. Any of those could address the customer's problem, and even in the late nineteenth century, all three options already existed. You wouldn't need to invent an entirely new technology; you would just need to figure out how to apply an existing technology to a new application. You would still have plenty of work to do—study the relative strengths and weaknesses of each option and select the most promising alternative—but at least you would be addressing the real customer problem.

The more you leverage from existing technology, the better your chance of success. Rarely will a disruptive innovation use entirely new technologies. It should leverage skills and technologies you already have, although you may need to add expertise in certain areas. Hiring consultants or specialized experts on temporary assignments can help, but your goal should be to grow and retain your core competencies in-house.

As for Steve Jobs, he was known for wandering unobtrusively through Apple stores to overhear what people were saying. He did not put a high priority on formal market research because he was acknowledged to have a one-in-a-million level of talent for understanding customer problems better than customers did. Most innovators won't have that level of talent, and it is unrealistic to expect it of them. Even Jobs did not get it right every time. Would he have priced the Lisa computer at $9,995 if he had talked to even a few customers?

You are not Steve Jobs. If you innovate based solely on your own intuition, your chance of success is much less than if you gather input from others: customers, noncustomers, colleagues, even competitors. Again, don't expect to be told the answer. Listen so you can truly understand the problems.

Let's do another thought experiment. Imagine it is the late 1990s and you are doing market research on portable MP3 players to help Apple decide how to enter the market. It is a market dominated by the Sony Walkman with players that use removable disks or even audio cassettes. In conversations with users, you may hear them say something along the lines of "I like being able to create my own playlists, but I can't put all the songs I want onto Sony's removable drive. Make a removable drive that can hold more songs."

That is not what you should do. The user says their problem is that they want their player to hold many more songs than it now can. But the user's idea for a solution—a removable storage device with more capacity—is uninspired. Spend a few minutes observing that user and you will learn even more: how they fumble through an awkward interface trying to find the song they want to play; how the player itself is too large to put in a pocket, and even if it could fit, one wrong bump would cause it to stop playing. And maybe the audio quality isn't all that great. With these inputs you will understand the real problems and may be able to invent a solution that even the user didn't imagine. (No, this is probably not the process Steve Jobs followed, but it is a good one for ordinary mortals to use.)

Apple's answer was the groundbreaking iPod. Not every company would be able to imagine an iPod. Even if they identified the same set of user problems, they might lack the expertise to envision a similar solution. Apple already had technologies to draw on: experience with high-capacity hard drives and flat-panel displays from their line of computers, and their existing iTunes app they could use to give customers easy access to many of the songs they would want. The iPod was a disruptive innovation that heavily leveraged technologies and skills Apple already had.

Those who advocate that innovation should be done in isolation aren't completely wrong. Actual innovation is best done in a quiet environment without interruptions. Once you have those user insights and fully understand the market opportunity, you will have what you need to innovate in an environment free of distractions, whether individually or in a small group.

Two other points are important. First, you might think innovation should be done by following a strict process, such as organizing formal group meetings to generate ideas, then ranking the ideas in priority order and assigning them to team members to investigate and report back. I suppose some companies do it that way, but in my experience, the best ideas originate when a few people are informally chatting around a coffee pot during a break or in the cafeteria over lunch. This kind of approach is great for coming up with ideas, most of which won't go anywhere. When a promising idea does emerge, innovators will need time to develop it in an environment free of distractions. The important point is to give people permission to innovate, not to make them follow a rigid process.

Companies staffed primarily with remote workers need to explore ways to fill this gap. This is one justification for a hybrid rather than a fully remote work environment. Just make sure everyone comes to the office on the same days to facilitate that interaction. For companies whose employees are geographically dispersed, regular online meetings with time reserved for innovation discussions can be an effective alternative. And if at all possible, get the team together live at least occasionally (annually or semiannually) in a neutral location away from distractions to conduct business planning. This kind of environment is great for inspiring innovation.

Another key point is to understand the difference between customer research and market research. In the former, your inputs come from people who currently use your product or service. In the latter, you go beyond those users to learn the needs of people in the target market who are not your customers. If your goal is to capture as much of the market as possible, you need to perform that expanded research.

SEGMENTING THE MARKET

Before you can decide how to best serve a market, you need to know what part of the market you want to serve. Most large markets consist of a variety of users with distinct needs. You can't satisfy everyone with your initial solution. Market segmentation is the process of dividing the market into logical groups of users who have similar needs.

You can segment a market in many ways. The automobile market, for example, can be segmented by such factors as customer income, sex, age, lifestyle, or geographic location. Customers in the snowy Northeast prefer four-wheel-drive SUVs, not the sporty convertibles popular in Southern California; soccer moms want more practical cars than teenage boys do.

The best way to segment a market will depend on your purpose. If your goal is to drive innovative product development, you will segment it differently than if your goal is to build a marketing campaign to attract new customers to an existing business. Most published guidance on market segmentation focuses on business-to-consumer markets such as the automotive example above. My experience is with the business-to-business (B2B) market, so that is what I will discuss. Similar concepts can be applied to research in consumer markets.

Refer back to Table 3.1 in chapter 3 to recall how we segmented the fiber-optic market for the DCA project. That table highlighted the key characteristics of customers in each segment and the differences across segments. It also made recommendations for the priority and action plan for each segment.

Table 3.1 is a simplified version of the more general template shown in Table 9.1. This template is a way to capture the important characteristics of each segment: customer needs, purchase decision process, key technologies, relevant industry standards, and example customers. The priority for each segment depends not only on those factors but also on how well the needs of a segment align with your ability to deliver a solution. These decisions are not absolute; they will differ by company. In the iPod example above, it would be easier for Apple to deliver an innovative solution than it would be for Procter & Gamble, which may sell products to the same target customers but does not have the necessary technical expertise.

It is also important to understand how the purchase decision is made. The final decision-maker may not be the most important person in the process. Often, that person will gather inputs from several others. For example, when a company needs to purchase a new set of servers for their IT network, the chief financial officer may be the one to make the purchase decision but will rely on guidance from the IT department for technical requirements, from operations management for software requirements, and from facilities engineering for space and power requirements. If you are exploring innovative ideas in server technology, you would want to talk with all decision-makers, not just the CFO.

Once you know the people you want to approach, you need to decide how to reach them. In the consumer world, web-based surveys and telephone interviews are commonly used. Online reviews posted by legitimate users on seller sites can also help you learn what customers like and dislike about your current product, as long as you are confident those reviews are real, not fake.

Methods for researching the B2B market are different. Whereas a consumer market may have a customer base in the millions, a business market may consist of anywhere from a few dozen to a few hundred customers. (Yes, I know companies like Microsoft or Apple have millions of business customers, but they are not typical of most business markets.) A better approach is to talk with representatives from target businesses directly, either in one-on-one sessions or in small groups.

	Market Segment #1	Market Segment #2	Market Segment #3
Relative Market Size	High/Medium/Low		
Market Driving Forces/ Customer Needs	• Describe the most important customer needs e.g. reliability, accuracy, ease of use, completeness of solution, etc.		
Purchase Decision Drivers	• Describe everyone who makes purchasing decisions and the factors that influence their decision.		
Key Technologies	• Describe the key technologies needed by each segment.		
Relevant Industry Standards or Legal Requirements	• List any industry standard or legal requirements that are important for this segment.		
Example Customers	• List names of example customers – company, job function, etc., as detailed as possible.		
Difficulty of meeting market needs	• Explain the most significant challenges for meeting market needs and the anticipated difficulty of delivering the solution.		
Conclusions	• Describe the actions to take to address this market segment: Act immediately? Monitor? Ignore?		
Priority	1	2	3

Table 9.1: Example template for B2B market segmentation

Online methods are common today. One approach is an online market research community. It is similar to a social media networking website but consists of a closed list of participants invited by a moderator. Participants (typically around fifty to seventy-five people) are drawn from the company's current customer base. The moderator posts survey questions and asks participants to comment. The entire community can see the responses and add their own thoughts, often generating a lively discussion. To encourage contributions, members usually receive at least a nominal reward for participating.

Another approach is a customer advisory board. This is a way for representatives from the company's C-suite to meet with senior executives from customers. Advisory boards (typically consisting of six to eight members) often meet quarterly, with in-person meetings once or twice per year and online meetings in the other quarters.

A customer advisory board should not be a decision-making group. Its purpose is for advisory board members to (1) share what they like and dislike about your company's products and services, (2) provide feedback on new products in development, and (3) offer insight into future market directions. Although members only come from current customers, they should be selected based on their ability to represent the needs of the entire market.

Online groups or executive-level advisory boards can be effective at getting general feedback from current customers, but there are times you need to reach further. At Agilent Technologies, we once wanted to get feedback on several possible options for a version of the DCA we were planning to launch into a new market. We did not currently serve that market, and we needed feedback from potential users to determine the best version of our planned product. Online sessions wouldn't be effective. We needed to get people together in a room for hands-on sessions with actual products.

We did not want to alert competitors to our plans, so we needed to assure that participants did not know they were talking to Agilent. Our marketing department recommended we use a variant of a technique known as a focus group.

A focus group is a way to collect qualitative data on a specific topic by pulling together a small group of participants in a neutral environment led by an outside facilitator. Focus groups have been used for everything from ranking which versions of a breakfast cereal people like better to which topics a political candidate should be prepared to discuss. Often, the participants know who is sponsoring the event and what they are looking for. Compared to interviewing people individually, focus groups add value by encouraging group discussion to bring a new level of insight.

Focus groups are controversial. They are expensive, they take time, and they rely on a small set of participants to accurately represent an entire market. One dominant personality in the meeting could skew the outcome away from the true feelings of the group. Especially in consumer-focused environments, focus groups have had a mixed success rate.

We understood those limitations and were ready to work within them. We started by hiring an outside company experienced in leading focus groups. They assigned our project to a skilled facilitator who had led focus groups many times before.

We decided to conduct focus groups in three major cities around the nation. We wanted participants to be a mix of both current customers and noncustomers. For customers, we collected recommendations from Agilent sales representatives in each region. For noncustomers we identified people who had published relevant papers or had established a credible industry presence. We gave the facilitator the name and contact information for each target participant. The facilitator made the phone calls to briefly describe the purpose of the session and extend an invitation to attend. He explained the anonymous sponsor would cover participants' local travel expenses plus provide a small cash reward as a thank-you gift. We limited attendance to eight participants at each location.

We wanted participants to test each of several design options and tell us which they liked best. Since we did not want them to know the sessions were being conducted by Agilent, we had our R&D team create versions of the product in generic packages with no indication of the manufacturer.

We held the sessions at commercial facilities designed for conducting focus groups. One wall of each meeting room had a large one-way mirror in front of a separate dark and soundproof room where several Agilent team members could observe the session anonymously. In addition to the facilitator, we asked an Agilent engineer who was unknown to the participants to conduct the technical portion of the session. We were careful to assure that they did not wear anything that would suggest they were from Agilent.

The sessions were successful and helped us select what proved to be a good option for the final product. Interestingly, at the end of each session, the facilitator asked the participants what company they thought was sponsoring the event. Almost everyone thought it was our largest competitor. A few guessed that one of the large Japanese manufacturers was trying to break into the market. Almost no one guessed Agilent. We did not correct their guesses, but when we introduced the product a few months later, the answer became obvious.

COMMUNICATING THE PLAN

Whenever I consider sharing text-filled tables such as those of Table 3.1 or Table 9.1 with the company's C-suite, I recall a quote from the 1984 motion picture *Amadeus*. After hearing a new opera, Emperor Joseph II said to Mozart, "Your work is ingenious! It's quality work. There are simply too many notes. Just cut a few and it will be perfect."

Don't get me wrong. The work a team does to segment markets is essential. That table should serve as one of the drivers for all project decisions. But its purpose is to aid the project team. A text-filled table is not the best way to communicate decisions once they are made. Whether you are pitching to a C-suite or to outside investors, you need to understand how they make decisions so you can tailor your presentation to their needs.

Most senior executives do not want to delve into all the technical details of your innovative project. Their objective is to assure that you have put together a compelling plan that has significant upside for the company and will fit within the planned budget. Their reaction to Table 3.1 would likely be "Too many words."

This chapter is not a treatise on everything you should know about pitching an idea to the C-suite. There is no single way to do it. You need to understand what your C-suite will expect. Will they want to drill into the technical details, or will they be more interested in market and revenue projections? In your prep work, talk to people who have made presentations to the C-suite before to learn what to expect.

Let's see how you could convert the text-filled Table 3.1 into a graphic that should resonate with senior leaders. Figure 9.2 illustrates one way to do it: a bubble chart showing the relative importance of each technology within the market.

The first thing to note is that the bubbles do not represent the market segments of Table 3.1, but rather the industry technologies from which those segments were derived. Internally derived market segments are useful for planning purposes, but always remember that industry technologies are what drive market trends. Those technologies are what you need to track as your project moves forward.

Figure 9.2 presents the major themes most leaders would want to know about the opportunity. It plots current market size against annual growth rate for each technology. The diameter of each circle shows the team's assessment of priority for that technology. Priorities are not set just by assessing market size and growth rate. They also need to take into account your ability to deliver a solution and the expense associated with doing so. Some segments may be lower in priority because although they may be growing rapidly, they don't align with your ability to deliver a solution. Most C-suite officers will not delve deeper than that, but you need to be prepared to do so if the question comes up.

You can often obtain estimates for market size and growth rate from reports published by industry consulting firms. Comprehensive reports can be expensive but include a level of detail that justifies their purchase price. Sometimes you can find what you need in summary reports posted on the consulting firm's website. Don't depend only on third-party analysis. You should certainly conduct original research through such actions as customer meetings and trade show visits, but in my experience, you will have more credibility if senior leaders know you have sought insight from third-party industry "experts."

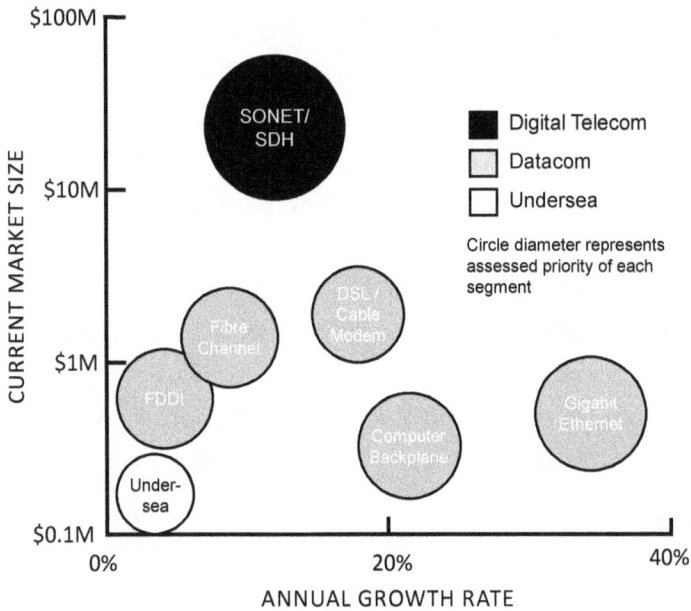

Figure 9.2: Bubble-chart version of Table 3.1 for use in presentations

Pitching an innovative idea to the C-suite is not the same as pitching it to outside investors. Although there are similarities, with outside investors you need to have a well-developed story before you make your pitch. With the C-suite, if you wait until then to share it with them, you are likely to be shot down. Avoid this risk by keeping senior leaders informed of your work from the outset. They are more likely to be supporters if they feel they have been able to help shape the plan. If they are not supporters, you will at least know that before having made a significant investment.

NEXT STEP

We have demonstrated throughout the book that innovation does not happen in some mythic place in the sky, in isolation—it is connected to the social dynamics of the work environment and the broader forces of industry. Accordingly, there is one more thing you need to do before you make your pitch. You need to align your innovative idea with the position in the life cycle your target market is in. Innovation targeted for a new, rapidly growing market is different from innovation targeted for a mature market. The next chapter covers this topic in detail.

Key Takeaways for Market-Driven Innovation

1. Large, market-leading companies can fall into the trap of believing that they know the market better than their customers do and don't need to conduct outside market research. This may cause them to miss important trends and expose parts of the market to attack by smaller companies.

2. Market research won't tell you what new products to introduce. Customers can tell you their problems, but their ideas for solutions will be unimaginative. It is your job to identify the true problems and come up with innovative ways to solve them.

3. The more an innovation can leverage technologies and expertise you already have, the better its chance of success.

4. Innovation doesn't need to be done by following a strict process. Sometimes the best ideas come during informal conversations over break or during lunch.

5. Market segmentation is the process of dividing a large, complex market into subsets of customers with similar needs. It is a tool that will help you decide the most attractive parts of the market to go after.

6. When you are ready to communicate your plan, look for ways to communicate your points succinctly. Visual graphics are superior to text-filled tables.

DISCUSSION QUESTIONS

1. Why do you think Ford stayed successful for twenty years selling essentially the same car for that entire period?

2. Apple was not the first company to introduce a portable MP3 player. Why do you think they were able to snatch market leadership away from Sony? What could Sony have done differently to retain its market leadership?

3. Think about the markets you currently serve. What are some ways those markets could be segmented into smaller groups of customers who have similar needs? How well do your thoughts align with how your company has already done this?

4. What are your thoughts on how to best conduct market research for the markets you currently serve? How would you make sure you got inputs both from current customers and from noncustomers you are trying to win?

5. Imagine you have what you think is an innovative idea you want to pitch to upper management in your company (either your direct manager or C-suite management). From what you know of your intended audience, what do you think would be the most important topics they would expect you to cover, and what would be least important?

Chapter 10

Innovation as a Function of Market Phase

The optimum strategy for innovation depends on the phase of the life cycle a market is in. A market in high-growth phase calls for a more aggressive approach toward innovation than one in a mature phase. A small start-up in a newly emerging market should take the most aggressive approach. A large multinational corporation that serves multiple markets should use a blended strategy to best serve differing needs.

Refer to Table 10.1 for the four primary market phases. I base this on the work of Theodore Levitt, who introduced the concept of market phases in his article "Exploit the Product Life Cycle" in the November 1965 issue of *Harvard Business Review*. My four phases align with his, and the table includes my recommended innovation priorities for each phase. Other authors define as many as six phases, but true to my role as a manager rather than an academic, I have kept the number to what I consider the most important four that give managers the insight they need to make informed decisions.

BUSINESS PHASE	EMERGING	GROWTH	MATURE	DECLINE
CHARACTERISTICS:	• Milestone-based strategy • Large cash investment not funded by revenue • Long breakeven time • High risk, high return	• Large dollar volume revenue • Excellent growth/profit • Large investment funded by profit • Rapid return on investment	• Stable revenue • Low investment level • Moderate return on investment • Investments focus on building customer relationships	• Declining revenue • Very low investment level • Low return on investment • Investments focus on most profitable market segments
OBJECTIVES:	• Answer technical and market opportunity questions before making major investments • Identify ideal customers • Establish company's reputation in new market • Pursue high ROI projects	• Grow revenue, profitability • Increase market share • Build organization that can deliver to customer demand • Keep track of disruptive innovation threats and respond as appropriate	• Maintain revenue, maximize profitability • Become customer's preferred supplier • Promote differential value • Watch for disruptive innovation threats	• Shut down money-losing businesses • Maximize profitability of remaining businesses • Move staff to other businesses • Look for new disruptive opportunities
INNOVATION PRIORITY:	Disruptive innovation to enter new markets	Incremental innovation to add new customer value	Incremental innovation to add customer value, improve profit	Redirect innovation to enter new markets

Table 10.1: Four key market phases and the innovation priorities for each

Emerging Phase. This is the phase in which disruptive innovation is used to enter a new market. Entering an emerging market requires considerable investment over an extended period. Projects in this phase are risky—that is why many more start-ups fail than succeed. Because of this, the expected return on an investment must also be large.

Customers in the emerging phase are a small set of early adopters. In the consumer world, they tend to be people who are enamored with technology and want to be the first to own anything new. In the business world, they are customers who are working at the leading edges of technology and are the first to need the new capability.

Emerging markets are the domain of start-ups, but when entered by an established company, they should be managed using the process described in chapter 7. The biggest threat for an established company is a senior leadership team that does not understand the concept of risk in a disruptive environment. They could well damage the career of any manager who does not deliver spectacular results every time.

Growth Phase. In this phase the market has moved past being of interest to only a few early adopters and has been accepted by the first users in the mainstream market—a transition consultant Geoffrey Moore calls "crossing the chasm." For the overhyped markets typical of high tech, this phase can turn into a tornado—a period of explosive growth in a market brimming with competitors.

The growth phase is characterized by a large influx of customers new to the market. For example, in today's market for electric vehicles, most purchasers are buying their first EV. Similarly, when railroads were purchasing diesel locomotives to replace their steam locomotives, the market was in the growth phase. Later, when they began purchasing new diesels to replace older diesels, it marked the transition to the mature phase.

Success in the growth phase depends on several factors. First, you need to develop strong relationships with your customers, so you are top of mind as their go-to source. Second, you need to build an organization that can deliver products quickly in a market where orders are growing rapidly. This means adding staff to build and sell those products and ordering enough lower-level parts to meet the expected demand. In the HP case study of chapter 3, soon after we launched the DCA, our biggest challenge became building products rapidly enough to meet demand. We would have grown even faster had we not lost orders to Tektronix because customers were unwilling to wait for the duration of our quoted lead times.

You need to have a strong sales organization during the growth phase. Maximizing the profitability of each individual sale is not your goal; you should be trying to get the most total profit dollars per month. Good salespeople have the hunter-killer instincts to do this. They do not care about the profitability of the deals they close; their objective (and how they are rewarded) is to maximize top-line revenue.

Product line management needs to work with those salespeople to make that happen. Pricing flexibility is important. If a few more points of discount make the difference between winning and losing a large deal, as long as you have the capacity to deliver and it still makes a profit, do it. You will have increased your total profit dollars for the business by doing so. (If you have limited production capacity, you should prioritize deal opportunities to get the most profit dollars per month.) Get all the profit dollars you can in the growth phase because you just might need them when your business transitions to the mature phase.

Innovation during the growth phase should focus on delivering incremental innovation that regularly provides new value to existing customers before competitors can do so. This is not the time to depend on disruptive innovation for growth. Why not? Because any time you introduce a disruption into the market, you change the landscape. Customers who have been buying their solutions from you discover they must now adopt something new, and that opens them up to alternatives. You do not want to give them that option.

Here is an example of how we once avoided that problem with the DCA. Recall that Sierra was the product we introduced to replace our original BudLight DCA. One of our key objectives for Sierra was to assure it was compatible with plug-in modules from BudLight. We also made sure the software that customers had written for their BudLight-based test systems would still work with Sierra. It might not work as efficiently as if they had taken advantage of Sierra's new features, but they would not be forced to write new code.

In contrast, when Tektronix introduced its BudLight competitor, it was not compatible with any previous product, not even their previous sampling scope. Adopting the Tektronix product would have forced customers to make a considerable investment in new test software, with no compelling benefits.

If Sierra had not been compatible with BudLight, the situation would have been different. In that case customers would have had to make a new investment either way, and they might well have decided to switch to Tektronix. By making Sierra an incremental rather than a disruptive innovation, we avoided that threat.

You do need to stay aware of potentially disruptive innovations from competitors. Market leaders don't need to be first to market with any disruptions, but they need to be ready to respond quickly if the threat arises.

Mature Phase. In this phase market demand levels off and may eventually begin to decline. You are no longer attracting large numbers of new customers; you are primarily selling to your existing customers. One of first things you discover is that you no longer need all those employees you hired and all those lower-level parts you ordered during the growth phase, and they are draining your profits.

Not every transition to the mature phase begins with disaster. Figure 10.1 shows three possible paths. In the first, the market grows continuously during its growth phase and transitions seamlessly to maturity. In the second, it enters a disastrous decline before recovering. In the third, the market never reaches critical mass and eventually ends.[33]

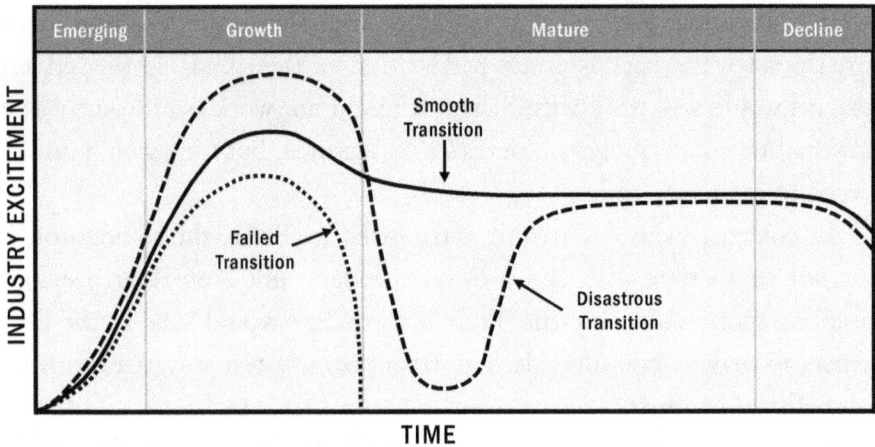

Figure 10.1: Three possible paths for market transitions
from one phase to the next

The gasoline-powered automobile industry is an example of a market that has been in its mature phase for decades. While some manufacturers went out of business during the transition from growth to maturity, the overall market did not go through a disastrous collapse (it may soon begin a slow decline as electric-powered automobiles gain share). The same cannot be said of the high-tech industry, which regularly ignores early warnings and watches with surprise as markets collapse, never to recover. Sad to say, I was once a prime example of a leader who ignored such warnings and faced disastrous consequences.

Here is that story. Recall the DCA-J case study of chapter 8. For several years, the original DCA business had been growing exponentially. In

---

[33] The vertical axis of this figure is labeled "Industry Excitement" rather than "Sales." Especially in emerging and growth markets, investment can be driven more by industry excitement or "hype" rather than actual need.

part because of its success, I had been promoted to the position of senior marketing manager for the entire division. The DCA was one of the many product lines in that division, but compared to others it was doing well, so I didn't spend a lot of time tracking it.

That changed when the fiber-optic market collapsed. I can't say we had no warning. During the height of our growth, Charlie Schaffer hired a product marketing engineer from outside Agilent who was a disciple of Geoffrey Moore. Donna Leever quickly realized we were in a tornado market. She warned the aftermath would likely be a market decline, but even she couldn't imagine the extent of the collapse. We had never been through such an experience before, and orders were still rolling in at near record numbers. Our financial analyst, Lincoln Turner, calculated that even if orders fell to half their current rate, we would still make double-digit net profit. The idea of telling our senior leaders we needed to prepare for an inevitable decline was unthinkable.

What we didn't realize was that customers who were desperate to receive products were placing orders with both us and with Tektronix. Their plans were to accept the orders they received first and cancel the others. At the time, Agilent's policy was to not charge customers a penalty for canceling orders, which made the eventual problem even worse.

There were other signs. Months before DCA orders plummeted, orders in other Agilent product lines selling to those same customers began showing noticeable declines. What did we do? Nothing. Like the proverbial ostrich, we buried our heads and ignored those signs. We continued to hire more staff and place large orders for lower-level parts—parts that would never be needed. One of the main reasons the DCA-J product line was unprofitable for over a year after its introduction was because of the write-offs we had to take on those unneeded parts for the earlier product. We had ignored the telltale signs, which meant massive layoffs and write-offs were inevitable.[34]

---

[34] This is a perfect example of where seeking guidance from a consultant like Geoffrey Moore could have reduced the extent of our disaster.

Now that you know what *not* to do in a mature market, let's discuss what you should do. First, don't think of a mature market as a dead end. Especially in high tech, a common belief is that if you are not chasing an emerging or a growth market, you are failing. That is a misguided belief. Mature markets should be the source of large profits even if they are not growing. Those profits are what fund investments in emerging markets. You need to maximize this revenue profitably, and you need to make sure the people assigned to this business receive recognition for the importance of their contributions.

Second, you should continue with incremental innovation. Again, think of the automobile industry. Although the market has been mature for decades, the automobiles you buy today are not the same as those from twenty or thirty years ago. Improvements such as antilock brakes, airbags, and advanced crash protection technologies have made cars safer. Improvements in reliability have allowed automobile manufacturers to extend warranties from yesterday's one-year period to today's three or more years. Incremental innovations such as these encourage existing customers to trade in their old cars to buy safer and more reliable new cars.

Even in a mature market, there is opportunity for a newcomer to succeed. The market for desktop and laptop computers has been mature for over a decade, but that has not stopped Carbon Systems from carving out a profitable growth business the Big 3 have been unable to replicate.

Third, you need to plan for the transition from growth to maturity before it happens. This change is predictable. There are always warning signs. Watch for changes across the entire landscape—customers who change buying habits or adjacent product lines that see a decline in orders. Sometimes simple math is all you need. The story in the HP case study of how the DCA sales manager calculated that if growth continued at its present rate for very much longer every man, woman, and child in the country would own a DCA should have been a wake-up call.

Once you accept the fact that change is on the horizon, you can make the appropriate plans. One approach is to spread out the risk. Use subcontractors for more of the work, not because it shifts the responsibility for layoffs to someone else (although it does that), but because contractors who serve multiple clients are better able to shift resources around as demand varies.

One warning: do not subcontract your core competencies. Keep that expertise in-house. Recall how locomotive manufacturer Alco partnered with GE for its first line of diesels. Alco subcontracted all sales and marketing to GE, so when GE ended the contract and introduced a competitive line of locomotives, Alco discovered it did not have any in-house experts who could take on sales and marketing for its own product line.

Another approach is to be careful about entering long-term contracts. This may seem counter to logic. Locking in long-term contracts to get the best prices on lower-level parts would seem to be the way to improve the profitability of every sale. But long-term contracts often restrict your ability to exit early, which is not a problem until the market collapses. Lock in contracts that give you plenty of parts when you need them without committing you to enormous numbers when you do not.

Decline Phase. There will come a time when the market has passed its prime and has begun to decline. The landline telephone market is one example; cable television is another. In this phase, your objective is to exit unprofitable businesses as quickly as possible and extract as much profit as you can from the remaining businesses.

Innovation in the decline phase falls into two categories. First are any incremental innovations you can do for a small investment that reduce cost or improve profitability of your current products. Don't worry about incremental innovation to add new customer value—customers won't care. How many people would even notice if you introduced a new landline telephone that incrementally improved the user experience?

The second form of innovation is that of disruptive innovation into new markets. In most cases, disruptive innovation projects are not the responsibility of existing businesses. Here, though, I am specifically referring to innovations that can take existing products into new, high-growth markets. The DCA-J of chapter 8 is an example. We took an existing product, added what could be considered incremental features, and redeployed the product to a new, fast-growing market. The result was a disruptive innovation that quickly became the market leader.

THE ROLE OF ACQUISITIONS IN INNOVATION

There are many reasons for a company to acquire or merge with another company. Most have little to do with innovation; their objective is to grow business through such avenues as market expansion, geographic expansion, cost reduction, or even by taking a competitor out of the market.

I will not cover those reasons here. My goal is to explore acquisitions done to obtain innovative technologies—typically when a large company buys a start-up. Technology acquisitions should be done during the emerging stage of a market life cycle. They rarely make sense in later stages because the priority then should be incremental, not disruptive, innovation.

Buying a small start-up is a strategy fraught with risks. It is so risky that many companies refuse to do it. A start-up is often satisfied with product performance that would never pass muster in a multinational corporation. The software may be full of bugs, the technical performance may not meet claims, and the hardware reliability may be questionable. Start-ups are willing to deal with those risks because they need to get a product onto the market quickly to start generating revenue. Their customers accept this risk because they want access to the start-up's advanced technology.

Customers of an established corporation have much higher expectations: bug-free software, performance that meets promises, and products that work reliably. If a large corporation puts a product onto the market that compromises those goals, it risks damaging its reputation.

One challenge for the acquiring company is the difficulty of performing sufficient due diligence to uncover all technical risks. A start-up will be reluctant to share full details of its technology unless the acquiring company signs a nondisclosure agreement. The acquiring company will not want to do that because if they later decide not to make the acquisition, the NDA may prevent them from inventing their own similar product. The level of due diligence the acquiring company conducts on the innovative technology may only be enough to make what seems to be a reasonable decision.

In my career I have only acquired one start-up for its innovative technology. As acquisitions go, it was a small purchase in the single-digit millions of dollars; it did not have a significant impact on our company's bottom line. But unfortunately, it did not go well. Here is the story.

Refer again to the DCA-J case study of chapter 8. That product was designed primarily for use by R&D engineers. While manufacturing customers built the DCA-J into complex production test systems, it was only because they had no alternative. They wanted a much simpler and less expensive product they could deploy in volume on their production lines. The start-up I wanted to acquire had developed just such a product.

There is no need to bore you with the details of the story other than to say it did not work out as planned. Instead, I will share the learnings that came out of that painful experience.

- **Do not make an acquisition without the full support of your team.** In my case, our marketing department was strongly in favor of the acquisition. They had been hearing from customers who were impressed that the start-up had designed a product specifically for manufacturing that sold for only half the price of ours. Marketing was adamant that if we did not acquire that start-up, our strongest competitor probably would. If the start-up's product was as good as it appeared, it could put half our business at risk.

  The R&D team, though, had little interest in it. They were confident they could invent something even better; they just needed the money to do it. My R&D manager claimed his team was fully engaged in other important projects and would have little time to help with the due diligence. He wanted me to divert the funds allocated for the acquisition to his team, which was impossible because they came from a different pool of money. When I decided to move forward anyway, I got only the bare minimum of

R&D support for the due diligence. I should have realized R&D's disinterest was a deal killer and stopped the project.

- **Do not rely on the start-up's view of the technology's readiness for introduction.** A start-up has a different view of what "ready for introduction" means. Bug-filled software and questionable technical performance may be acceptable to a start-up's customers, but not to those of an established multinational corporation.

- **Build a large internal investment into the plan so your team can help with development after the acquisition is complete.** You made the acquisition to obtain the technology. You should not expect the start-up to know how to introduce a product that meets all your company's requirements for shipment. Your team needs to help the start-up resolve problems such as software bugs or hardware reliability. Manage this project as diligently as you would an internal project.

- **Consider the alternative of investing in the start-up rather than buying it outright.** This is the corporate venture capital approach. It gives you time to assess the true value of the technology without locking you into an expensive commitment. You would still have first right of refusal to buy the start-up down the road after the technology has been proven. This is the approach I should have taken, but Agilent had no experience with it and viewed it as a foreign concept.

- **Your first acquisition will probably not go well. Use what you learn to get better at it.** It takes practice to get good at playing a professional sport; why should you think it will be any easier to get good at acquisitions? Give yourself an edge by getting help from people who have done it before. Ideally, the corporation will have an experienced M&A team that can help individual organizations who are going through it for the first time. Conduct a postmortem after every acquisition to capture ways to improve in the future.

Part of embracing market-phase-attuned innovation is accepting dynamics beyond your control. Not long after I made the acquisition, the world economy collapsed in the Great Recession of 2008. The start-up's technology was still far from ready, and Agilent needed to make drastic cuts. I was ordered to close the start-up and lay the people off, giving them Agilent's standard severance package. We never got the chance to prove the value of that promising technology.

This acquisition unquestionably failed to meet its objective, but I can make a case that it delivered a powerful benefit to the DCA business. With just a small total investment, we eliminated a serious competitive threat to a multimillion-dollar business and prevented anyone else from acquiring that product. This, however, was never my intent. If I had made such a proposal to our C-suite, they would never have approved the plan. While this acquisition did not benefit my career, it quite possibly prevented a competitor from stealing much of our business. Such are the vagaries of life as a senior manager in a multinational corporation.

Key Takeaways for Innovation as a Function of Market Phase

1. The priorities for innovation depend on the phase of the life cycle a market is in:
   - In an emerging market, disruptive innovation should be the priority.
   - In a growth market, incremental innovation that adds customer value should be the priority.
   - In a mature market, incremental innovation that improves profitability or adds customer value should be the priority.
   - In a declining market, incremental innovation to improve profitability or disruptive innovation to enter a new market should be the priority.
2. Acquiring a small start-up for its innovative technology is fraught with risks. Consider making an investment to help fund the start-up rather than buying it outright, with the option to acquire it later.

DISCUSSION QUESTIONS

1. Geoffrey Moore, in his book *Inside the Tornado,* describes five market phases based on the maturity level of users. From earliest to latest, he classifies them as techies, visionaries, pragmatists, conservatives, and skeptics. How would you map these five phases onto the four phases described in this chapter?

2. Why is disruptive innovation not recommended during the growth phase of a market? Are there some circumstances where it would make sense?

3. What are some signs to monitor to help identify when a market is moving from the growth phase into the mature phase?

4. What are some ideas for how a company can remain profitable in a market in the decline phase?

5. What are some of the risks associated with acquiring a small start-up for its innovative technology? What are some ideas for how to reduce those risks?

Chapter 11

Driving Innovation by Influencing the Industry

S o far, we have discussed innovation in terms of how to drive it within your organization: its management structure, employee training, incentives, alignment with markets, and alignment with the phase an industry is in. In this chapter we explore how to drive an innovation by influencing the broader industry. To illustrate how this can be done, we will again start with a story.

AN INDUSTRY IN TURMOIL

In 1986, electronics manufacturing was in turmoil. For years, electronic products had been built using printed circuit boards whose components were attached by way of a process known as through-hole technology. Those components had wire leads that extended from their bodies. Specialized machines picked up the components, inserted those leads into holes drilled into the printed circuit board, and then soldered them in place. This legacy technology had worked well for the electronic products of the 1960s and 1970s, but it was insufficient for the increasingly complex designs of such products as laptop computers, laser printers, portable audio players, televisions, and industrial equipment of the 1980s.

Early in the decade, a disruptive innovation known as surface mount technology had emerged as a candidate to replace through-hole technology (Hinch 1988). Surface mount components are smaller and less expensive than through-hole components (see Figure 11.1). They do not rely on holes in the board for attachment. Instead, they are soldered directly to the surface of the printed circuit board. Surface mount components can be placed on both sides of the board, increasing capacity by as much as four times.

Figure 11.1: Printed circuit board assembled with both surface mount and through hole components. The tall tubular components are capacitors that use through-hole technology, as is the one large voltage regulator integrated circuit at the top left of the photo and the large connector at the top right. All other components are surface mounted (photo by the author)

Companies worldwide rushed to embrace the new technology. Component manufacturers designed entirely new families of components expressly for surface mounting. Equipment manufacturers designed elaborate machines that could take those components and mount them onto printed circuit boards. End users began designing products to take advantage of this new technology. Yet even after five years of intense effort throughout the industry, surface mount technology was languishing. In consultant Geoffrey Moore's words, it had failed to cross the chasm.

The most severe problem was that every component manufacturer had adopted its own set of standards. Components from one manufacturer were rarely interchangeable with those from another. Although industry organizations had been working on standards, they faced a difficult challenge. Component manufacturers would protest and possibly take legal action if standards were written in ways that excluded their products, so the resulting standards had to be extremely broad. It was impossible for one printed circuit board design to work reliably with components from different manufacturers.

End users were reluctant to create designs using components that might soon become obsolete, and manufacturers were reluctant to gear up for high-volume manufacturing of components that might never see orders. A hot topic in the trade press was whether surface mount technology would ever enter the mainstream.

Hewlett-Packard Company was caught squarely in the middle of this turmoil. We were one of the industry's largest users of electronic components across such diverse product lines as computers, printers, calculators, test equipment, and medical diagnostic equipment. We needed surface mount technology more urgently than most companies did.

I was right in the middle of this dilemma. I had been recruited to a new position in HP Corporate Manufacturing to lead the development of a common set of surface mount standards across all HP divisions. The turmoil in the components industry was a major obstacle.

I pulled together a team of technical experts from around the company to draft HP's standards for all aspects of the technology. For components, we selected a set of existing commercial designs we knew would work for us, but we had a problem. Many of these components were only available from one or two companies. We did not want to be locked into families of components that were not widely available. We had tried to get our selections adopted as industry standards but were unsuccessful because some manufacturers objected (Bennett 1985).

Progress for surface mount technology throughout the industry was at a standstill, and I decided the only solution was for HP to take matters into our own hands. We were a user, not a component manufacturer. Although we could not set industry standards, we could tell the industry the full specifications of the components we were going to buy. Manufacturers that met our standards would get our business. Those that didn't would not. I hoped we were a big enough customer to get the industry's attention.

I was a member of the board of directors for SMTA, the industry trade association for surface mount technology. I knew John Tuck, the editor of the widely respected trade magazine *Circuits Manufacturing*. He had been one of the most vocal journalists lamenting the sorry state of surface mount manufacturing. When I told him my idea, he encouraged me to write an article. He would see that it got published.

My article ran in the March 1986 issue of the magazine. It spelled out the complete specifications of HP's preferred components, written as if it were a true industry standard. I had expected the article to be buried near the back of the magazine, but John made it the lead story. He commissioned an artist to create a magazine cover showing an army of Lilliputians trying to drive a square peg through a round hole, and he highlighted this quote from my text: "It is little wonder that surface mount technology has failed to fulfill the optimistic projections of the last few years. The time has come to force the issue."

As soon as the magazine was printed, I was flooded with calls and letters. End users thanked me for taking the initiative. Many agreed to adopt the HP standards. Component manufacturers reacted positively, although a few tried to steer us toward their current products. When that failed, they realized they would need to develop product families that met our requirements. This one article helped break the surface mount logjam and drive the technology forward. Within three years surface mount technology had become dominant in all major segments of the electronics industry.

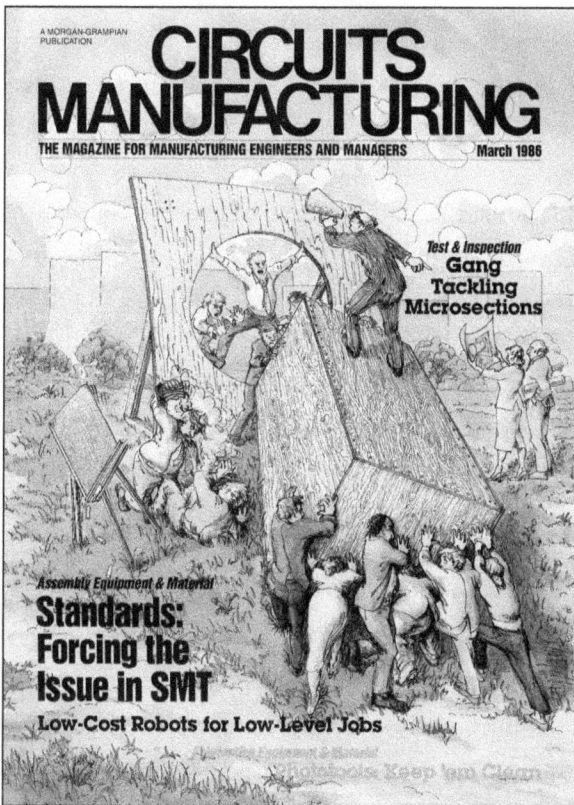

Figure 11.2: March 1986 issue of *Circuits Manufacturing*, the issue that drove adoption of surface mount component standards (author's collection)

This was an eye-opening experience for me. Previously, I had been deluged by HP engineers who questioned why I was taking such an active role in the outside industry: "Why are you telling everyone something that should be an HP secret?" This experience convinced me that influencing the industry was important. Not every decision a company makes should remain a trade secret.

WHY DRIVE INDUSTRY STANDARDIZATION?

Companies that launch disruptive innovations often feel they need to protect those innovations from outside competition. After investing so much time and money to develop and launch the product, they do not want competitors to get a free ride into the new market. Such an approach never works out. Customers want choice. They do not want to be locked into a single solution.

The classic example from the distant 1970s was when two competing technologies for consumer video cassette recorders burst onto the market. Sony launched the Betamax format in 1975. A year later, JVC introduced the VHS format.

JVC believed in the value of an open standard. They lined up Matsushita, Mitsubishi, Hitachi, and Sharp to support the format. Low-cost VHS players from those manufacturers soon flooded the market. Sony, knowing its higher-quality video offered a superior viewing experience, kept prices and licensing fees high in an attempt to increase profit. It was yet another example of a company assuming that better performance will always lead to more sales. But when a competitive product has enough performance to meet the needs, customers will base their decision on other criteria—in this case price. Betamax lost the format war after only a few years.

A recent example is that of the charging cable for electric vehicles. Tesla created one plug format, now known as the North American Charging Standard (NACS). A consortium of seven manufacturers—Audi, BMW, Daimler, Ford, General Motors, Porsche, and Volkswagen—created a different format known as the Combined Charging System (CCS). Neither works directly with the other, although adaptors provide rudimentary compatibility.

Tesla has more than twice as many chargers as CCS throughout North America, and they work over 90 percent of the time. CCS chargers are much less reliable. One widely cited study found that CCS chargers were out of service more than 25 percent of the time, often sitting unrepaired for weeks or months (Rempel et. al. 2022). The sorry state of CCS chargers has been a major source of customer complaints.

Tesla could have turned this into a selling point to give their EVs a leg up on other manufacturers. Instead, they took the opposite approach. They realized that standardizing on one plug format would benefit the entire industry by removing an obstacle that made customers hesitant to buy. That would help Tesla grow even faster, so they promoted theirs as an open standard. Given that Tesla chargers were widely available and highly reliable, other manufacturers had little choice. They all adopted Tesla's NACS standard as of their 2025 models.

Tesla did not do this out of a sense of altruism. They spun off their charging network as a separate company that will generate profit from every car that plugs into a Tesla charger—the next Standard Oil in the making?

Not every disruptive innovation will have that kind of opportunity to influence the industry. But if it does, put all your energy into promoting it. Here are four things to do that will give you the best chance of success:

- **Start by getting support from your own company's senior leaders.** Driving industry standards takes time. You will need to do more than just attend standards committee meetings. Volunteer to serve actively on those committees (as chairperson, if possible), speak at conferences, and travel as necessary to exert influence. Your company's senior leaders need to understand and support this commitment.
- **Establish your presence in the industry before attempting to drive a standard.** The more visible you are, the better your chances of successfully promoting your innovative idea. When *Circuits Manufacturing* published my magazine article on surface mount component standards, I had already established a reputation in the industry. I had joined the trade association SMTA two years earlier and volunteered to teach SMTA-sponsored classes on surface mount technology at venues around the world. I was soon invited to serve on the SMTA board of directors, which led to invitations to be a featured speaker at conferences. The component standards I proposed were readily accepted because they did not come from an unknown player with questionable credibility.
- **Be sure your proposed standard is better than the alternatives.** Simply floating your idea out as a proposed standard will not be sufficient if you cannot prove its value beyond all other

alternatives. Imagine how the electric vehicle market would be different today if the consortium of manufacturers behind the CCS connector had put the necessary effort into assuring it was as reliable and as ubiquitous as Tesla's. With seven manufacturers behind it to Tesla's one, the industry would have undoubtedly adopted CCS, not NACS, as the North American industry standard.

- **Build a coalition of support across the industry.** No matter how good your proposal, it will fail if it doesn't have broad support from others outside your company. Take the time to gain that support and be ready to modify your proposal as necessary. That doesn't mean you need to compromise your own plans. Recall from chapter 3 that when I wrote the industry standard for testing fiber-optic transmitters, I accepted a less rigorous approach that was backed by several companies. That didn't stop my team from introducing a product that far exceeded the standard and won the market by a large margin.

Finally, I will address the question some readers may have about the validity of my premise that market leaders will not be successful if they do not open a disruptive innovation to competitors. In the mobile phone industry, Apple has never opened its iPhone technology to competitors. It has coexisted with Android for well over a decade. Does that not disprove the theory?

No. Remember, the critical factor is that customers must have a choice. It does not mean competitors must be allowed to copy the leader's disruptive innovation. For the mobile phone market, customers have that choice. Both phones operate on the same cellular telephone network. Both phones talk to each other over that network. Both are true smartphones that can run versions of the same apps. There is little worry that one product will shut down the way Sony shut down the Betamax video cassette recorder product line and left those customers with cartridge recordings they could no longer play. And similar to how EMD allowed Alco to survive in the locomotive business to avoid being declared a monopoly, Apple has not tried to shut down Android. Sometimes détente is the best answer for all players.

CONCLUSION

Once you have mastered the fundamentals of innovation and set up your corporate environment to optimize innovative potential, you can potentially do something really exciting—begin to influence the course of an industry. But sound knowledge of the nature of corporate competition is absolutely necessary to making the innovation, no matter how brilliant, feel intuitively superior to the end user. You need to put as much attention into promoting your innovation as you did in inventing it.

Key Takeaways for Influencing the Industry

1. Companies that launch a disruptive innovation often feel they must protect their innovation from outside competition. That rarely is successful.
2. The consumer videotape recorder war in the 1970s demonstrates why an open standard will win against a closed standard. Customers in an emerging market want choices; they don't want to be locked in to a single supplier.
3. Tesla's decision to open their charging cable plug format to the industry is an example of how promoting an open standard is critical to the success of an entire market. Tesla will benefit more than if they had kept their plug format proprietary.
4. Before proposing a standard, establish your presence in the industry so it doesn't appear to come from an industry unknown.
5. Build a coalition of support for your idea across the industry.
6. Success is not guaranteed just by the number of companies that support a standard. Customers must believe it is better than any of the alternatives.

DISCUSSION QUESTIONS

1. Hewlett-Packard was able to drive the standardization of surface mount components even without the backing of industry standards organizations. What are three things that made this possible?

2. Tesla's decision to offer their charging cable plug format to the industry as an open standard had both positive and negative impacts for Tesla. Describe some of the positive and negative factors. In your opinion, do the positives outweigh the negatives?

3. What are some other examples of times when a company drove industry standardization in a way that clearly benefited that company?

4. During which phases of the market life cycle does it make sense to drive industry standardization?

5. Do you think it is possible to drive the adoption of an industry standard without being well known in the industry?

Chapter 12

Innovation in the Digital Age

I n my introduction to this book, I made the claim that the concepts behind innovation transcend the world's entry into the Digital Age. Now that we understand those concepts, it is time to justify that premise.

Predicting the future is hard. If it were easy, everyone would do it. So it is with a certain amount of trepidation that I begin the discussion of innovation in the Digital Age—a term coined to describe today's age of technology. The Digital Age emerged in earnest with the widespread adoption of personal computers in the early 1980s and became dominant when the internet entered mainstream life at the turn of the twenty-first century. It has transformed the economy from one of the traditional industrial age to one based on information technology. Microsoft CEO Satya Nadella has called it "a generational shift in our economy and society."

Digital technology permeates both our work life and our leisure life, from smart factories to smart cities to an explosion of internet-connected devices in our homes. Digital technology has made it possible for a physician to diagnose a medical problem without the patient ever needing to visit an office, for passengers to ride in taxis with no drivers, and for teams to collaborate while dispersed widely across the world. Some business teams have worked together for years without ever having met in person.

The Digital Age has many components, but for innovation, the one that has generated the most excitement is a tool known as generative artificial intelligence or "GenAI." The concept behind GenAI is to draw from vast quantities of stored data to answer requests posed by users. The data stored in the GenAI model's memory may not contain an exact answer to the user's request, but the tool takes what it knows and uses it to synthesize what it thinks is a statistically probable answer. Whether this answer is valid depends on the quality of the stored data and the ability of the model to glean accurate information from it.

Before the GenAI tool can provide answers, it must first collect and store an enormous amount of data. Tools intended for general use draw source material from many areas, including most of the publicly available data on the internet: Wikipedia, books, periodicals, online technical reports, images, code, audio recordings, and even social media sites such as Facebook and Instagram. The AI tool compresses the reams of raw data by encoding it into a more efficient form known as a large language model (LLM) that it prepares for use through a process known as *training*. Then, in response to a request entered by a user (known as a *prompt*), the model decodes what it needs from this training data to create an answer. This is not simply a replay of the original data; it is a new output constructed from that data and presented as human-language-like text or images.

Examples of GenAI models designed for general use include ChatGPT, Microsoft Copilot, Claude, and Google Gemini (formerly Bard). Other models have been designed for specific purposes. These contain more comprehensive data sets covering their intended applications without the clutter of a vast array of other data. Jasper AI is an example of a model designed for marketing purposes; AlphaCode and GitHub Copilot are two for software programmers.

GenAI is early in the growth phase of its market life cycle—the time when usage is exploding, with many people using it for the first time. As with any new technology, not everyone agrees on its value. Three camps have emerged. The first sees artificial intelligence as a breakthrough that will have tremendous positive value for everyone. The second sees it as an existential threat to human existence, stealing jobs and dominating our lives. A third sees it as an overhyped phenomenon that will eventually collapse the way so many disruptive technologies have done in the past.

People are still learning what GenAI can and cannot do. Some imagine it as the omniscient and threatening robot Hal from the movie *2001: A Space Odyssey*. Others see it as no more than a "stochastic parrot" that merely repeats what it was taught with no ability to reason (Bender et. al. 2021).

We are not here to discuss every possible way GenAI could impact society. Our goal is to explore its potential for use in innovation. As a first step, let's understand the power and limitations of today's GenAI by considering the following analogy:

Imagine you are a person who has a photographic memory. You can remember every detail of everything you have ever read. Your friends know this and come to you for answers to their questions. One day, a friend who is writing a term paper asks you what the gross domestic product (GDP) of Bolivia was in 1953. You have not read anything with that exact answer, but you did read a World Bank report that said the GDP across all of South America was $35 billion in 1950 and $50 billion in 1955. You also read a Wikipedia entry that said Bolivia was one of the poorest South American countries, accounting for less than 2 percent of the total South American economy in 1955. You take that information and process it to come up with your best guess for Bolivia's GDP in 1953. You pronounce it to be $632 million. (We will not explain here the math behind your number or why you decided to use only those two publications as your data sources; those are entire topics on their own.)

There are several potential problems with that answer. First, it is only an estimate that may be nowhere near reality, but it is such a precise number that it comes across as absolute. Second, not everything you have read may be accurate, and some of it may have been deliberately misleading. Unless you have some way of moderating your memory to remove inaccurate or obsolete entries, you will give incorrect answers. Third, many of the things you have read are copyrighted materials. If you quote that material in a format anywhere near verbatim, you will be in violation of copyright. Fourth, users will not know what source material you used or how you made your calculation, so they will not know how much to trust you. Finally, you can only provide answers based on what you know. If another person who has photographic memory but has read different publications were asked the same question, they would come up with a different answer.

This last point is especially important for innovation. You may have a photographic memory, but you can only provide answers based on what you know. Suppose a friend came to you and said, "Give me an innovative idea for replacing jet aircraft with another form of transportation. It does not need to be any faster than a commercial jet but should be half the cost and twice as friendly to the environment."

You would search through your memory banks and recall that you read about how SpaceX proposed an underground vacuum tunnel called Hyperloop to transport people between Los Angeles and San Francisco using capsules propelled by magnetic levitation. You have also read about companies working on suborbital transport systems. So you offer those two options as innovative alternatives. But they are not truly original because companies have already been working on them—that is how you know about them. If your friend decided to start working on your "innovative" proposals, they would already be behind the competition. You are not able to propose something truly innovative like a Star Trek transporter because there is no technology like that in your knowledge base. You know lots of things, but your expertise does not include a way to expand beyond that to propose a creatively disruptive innovation. This is the story of today's GenAI.

Actual GenAI is somewhat more complicated than this simplified explanation. There are various approaches to modeling. One widely used approach is known as a transformer-based model, introduced by researchers at Google in 2017. In this approach, the model does not store the original data directly; it converts the data into snippets of numbers called tokens. When a user enters a prompt, the model also breaks that request into numerical tokens. The model uses its predictive capability to guess what tokens stored in memory are most closely associated with the tokens in the prompt. It then constructs its answer from those predicted tokens, putting them into human-language-like sentences.

How does the model decide what tokens in its memory are closest to those in the prompt? It uses the data it was trained on. This is why GenAI models trained on different data sources give different answers. Even a single model will give a different answer to the same question at different times because its prediction algorithm takes into account more than just the raw probability of adjacent tokens; it deliberately introduces a certain amount of "humanlike" uncertainty into the process.

Since its answers are based on predictions that may not always be accurate, GenAI must be used with care. It is well suited for low-risk tasks such as creative arts, gaming, and summarizing lengthy documents—actions that can be reviewed by a human and edited as necessary. It is also good at improving the readability of draft documents and can even compose an initial draft when prompted.

Schoolteachers are legitimately concerned that GenAI will allow students to create entire term papers without ever understanding the subject. But attempts by universities to use GenAI tools to detect student submissions that were created by AI have shown mixed results. Students whose original writing too closely matches that of a GenAI tool such as ChatGPT have received failing grades on the unproven assumption they used AI to write their papers. Given the serious implications of false positives, experts agree that it is irresponsible to make decisions on the integrity of a student's writing based solely on an AI-detector's assessment (Hale 2025).

One example of a good use of GenAI is a chatbot that replaces humans for first-level customer support. When a user has a problem, they log onto the company's website and submit their request via chat. The response the chatbot returns makes it seem as if the user is conversing with a live human, but it is really an AI tool trained on the product documentation. As long as the user's request is related to that training, the AI tool will provide an accurate answer. If it is not, the chatbot should escalate the request to a second-level human support resource. While chatbots may occasionally get things wrong, such problems are usually no more than frustrating to the customer and are little different from what an entry-level support person might have told them.

GenAI can be problematic when used for serious, potentially life-changing purposes. One concern is that it can produce *hallucinations*—incorrect or nonsensical answers that seem reasonable. These can occur for several reasons. The prompt submitted by the user may have been ambiguous, or the model's training data may be inaccurate. Models that contain either too much data or too little data can produce invalid answers. Too much data can cause the model to "overthink" the answer. Too little data can cause it to make bold predictions based on insufficient information. Some models try to address this risk by being designed to reduce the chance of making predictions that are too extreme.

GenAI can still be useful in such situations as long as the user understands the risks and takes steps to minimize them. In a situation where accuracy is essential (for example, when analyzing medical images of patients for possible cancer), don't use a GenAI model trained on a vast array of data across many topics. Restrict the training data to what is relevant to the task at hand and be sure the model has been given plenty of data that is updated regularly. Remove data that has been determined to be erroneous or obsolete. GenAI's output should always be reviewed by a human who is knowledgeable on the topic.

Data aging is a serious concern in applications where knowledge is moving rapidly. The field of medicine is an example. If the LLM is populated with data that is years or decades old, it may not provide answers that align with current best practices. If the LLM is populated with a mix of new and old data, the outcome could be even worse. The model may generate a result that sounds logical but is far from accurate to today's standards. The risk of data aging is especially concerning for organizations that manage their own GenAI models. The organization must maintain a staff dedicated to keeping their customized LLM updated with current knowledge in the industry. Such a responsibility may well be outside the skill or financial ability of a small company to manage.

AI raises potential legal questions that anyone using it needs to be aware of. First is the question of copyright protection. LLMs draw from vast arrays of source materials, many of which are copyrighted. Reproduction of those copyrighted materials could constitute infringement unless justified by the "fair use" exception in the United States or "transient or incidental copying" exceptions in the European Union.

AI Wars

As this book was going to press, Chinese startup DeepSeek introduced what is being hailed as a disruptive innovation in GenAI, DeepSeek-R1. While the world is still discovering everything it has to offer, here are some things we already know.

First, DeepSeek does not fundamentally change how GenAI works. Like other tools such as ChatGPT and Claude, it stores data in a large language model. While its algorithms for extracting predictions from that LLM are different, there is no evidence they are inherently better than others. The discussion in this chapter about how GenAI works applies equally to DeepSeek as to other GenAI tools.

DeepSeek evidently delivers results faster but with less precision than other tools, on the assumption this is "good enough" for many users. It appears to be another example of the concept introduced in chapter 3 that better accuracy does not always lead to more sales. If a competitive product is good enough, users will make decisions based on other factors.

One of those factors is that DeepSeek claims to be far less expensive to deploy and use than other tools, and it needs far less energy to do so. While China appears to have vastly understated the cost to develop DeepSeek, it does appear to be less expensive to use. It is based on low-end versions of AI processors from Nvidia, not the high-performance versions used by others that the US government has prohibited from sale to China. DeepSeek also claims to use a powerful new technology known as agentic AI that allows it to improve its predictions by learning from past results without needing new sources of data. Other models have agentic AI capabilities, so DeepSeek is not unique.

DeepSeek's most lasting impact will probably be to accelerate the development of faster, less expensive ways to deliver GenAI. Today's market leaders must react. As this book has shown, market leaders do not need to be the first to introduce a disruptive innovation, but they need to act quickly when a competitor does so. Customers prefer to source their solution from a company they know and trust, and the China connection raises a potential security risk. Users have also found DeepSeek to have a notable China bias in its results.

DeepSeek is only the first of what will undoubtedly be numerous new entries into the market. The war to accelerate the expansion of AI has only just begun.

The question of copyright protection also applies to content generated by the AI system. In 2020, the European Parliament stated that works created by an AI system are not eligible for copyright protection because they are not created by a human. The US Copyright Office made a similar determination in a statement issued in March 2023. Works created by AI are not protected by copyright unless a human had creative control and "actually formed" the traditional elements of authorship (US Copyright Office 2023).

Another legal issue, particularly relevant in the EU, is the question of personal data protection when using GenAI. My son Greg is the CTO of a firm in the UK that delivers AI-assisted services to companies in the European financial industry. One of their major concerns is to assure that they adhere to the privacy and security requirements specified in the EU General Data Protection Regulation (GDPR). The GDPR requirements are so strict that the financial industry prohibits the use of any GenAI tool whose LLM is stored on servers located in the United States. This excludes the use of models like ChatGPT or Gemini. Greg has had success using a proprietary tool on Amazon Web Services.

Despite the known risks of GenAI, it is here to stay. At the time of this writing, we are in the early stages of the AI tornado. While much of the excitement is built around unrealistic hype promoted by those in the business to make a profit from it, GenAI is already being used extensively across a wide range of industries. A likely scenario is one in which AI performs the predictable tasks, leaving humans to focus on the true creative tasks. In that sense it is similar to the advances technology has made in the past. Before the rise of the handheld calculator, everyone did math using pencil and paper. Educators feared an entire generation of children would never learn how to add or subtract, leading the world to ruin. In reality, calculators made it easier for mathematically challenged individuals to do arithmetic more often than they ever did by hand. The same can be said for the rise of personal computers—people who would never have written a letter by hand found their calling writing text and emails on word processors. The rise of the internet made it possible for people to perform research far more quickly than by going to a library. And yes, all of these advancements came with problems—children hooked on computer games, adults hooked on questionable internet searches, trolls who would never have written and mailed a letter by hand suffering from diarrhea of the fingers, posting trash comments on social media.

AI will undoubtedly have a similar level of value and threat. We are already seeing the impact of realistic deepfake images, videos, and audio recordings that portray a person as doing or saying something they never did. In the short term, deepfakes may well undermine trust in news reporting, not only because of questions about whether a photo or video is real, but also because it allows perpetrators to claim the evidence against them is fake. It is interesting to note that companies are already developing AI-based detection systems that can flag fakes when they appear. I am hopeful society will eventually come to terms with the negatives of AI and be able to deal with them effectively.

FACING THE FUTURE

Whatever the future of AI, innovation will always be critical to success in the business world. And the most effective way to carry out innovation will remain embracing a philosophy of simplification grounded in high-contrast management—which starts with people. Managers must understand how to cultivate innovation in their organizations and how to encourage it with their employees. Do senior leaders provide the support necessary to innovate successfully, or are they hampered by the curse of the corporate business model? Are they trainable? If so, engage those leaders in ways that encourage them to contribute. If not, it may be time to look for employment at a more innovative company.

Senior management support is essential, but it is not sufficient. You, your peers, and every employee in the organization should believe innovation is possible and encouraged. The lessons shared in this book can help you train your team on how to drive it successfully. Now it is up to you to put those lessons into practice. May success be with you.

Key Takeaways for Innovation in the Digital Age

1. The Digital Age emerged with the rise of personal computers in the 1980s and became dominant with widespread adoption of the internet at the turn of the twenty-first century.
2. The technology known as GenAI offers considerable value to innovation but also comes with risks.
3. A key component of GenAI is a database known as an LLM. The user posts requests known as prompts to the GenAI model, and it provides what it thinks are statistically probable answers.
4. The quality of the answers depends on the quality of the data stored in the LLM and the ability of the model to extract accurate answers from it.
5. GenAI can be good at providing answers that are within the bounds of its stored data, but it does not have a true ability to reason. It cannot provide innovative answers that are outside the limits of its database.
6. GenAI is best used for routine tasks that it can perform accurately and more rapidly than a human. This frees up humans to take on true creative tasks that are outside the GenAI tool's ability.

DISCUSSION QUESTIONS

1. How is generative AI different from traditional online search tools such as Google or Firefox?
2. Google has begun adding an AI-based answer to the top of search results for its queries. What impact do you think this will have on a person's use of search results?

3. What are your thoughts on how GenAI can be used effectively in the innovation process? What do you see as the biggest concerns with using it?

4. What job types do you expect to be most impacted by GenAI? What job types will be least impacted?

5. The widespread emergence of deepfakes has cast a cloud over the believability of information retrieved from the internet. How do you see this affecting society in the coming years? What can be done to reduce the negative impact of deepfakes?

Acknowledgments

This book is the product of insights I have gathered over many years from many people. I first started thinking about writing it more than a dozen years ago. I even drafted a few chapters but had to put the project on hold when I became president of the TeamLogic IT franchise in Santa Rosa, California. Presidents of newly formed start-ups have little time to devote to writing books.

The idea sat untouched until one day during a consulting session I had with David Cook of Carbon Systems. He mentioned that a customer had encouraged him to learn more about innovation. He asked me if I knew anything about the subject. That was the trigger for me to dredge up and share with him my early draft chapters. After reading them, Dave strongly encouraged me to finish the whole book. So my first thanks go to Dave for lighting a fire under me to do just that.

Case studies are a core element of this book, and the people who delivered on those projects deserve special recognition. Hewlett-Packard's innovation in digital communication analyzers is one of the book's central themes. I draw on elements of that case study in several chapters to illustrate concepts of innovation. I thank the team I worked with to create that groundbreaking product line: my marketing counterpart, Charlie

Schaffer; my financial analyst, Lincoln Turner; project managers Mike Karin and Chris Miller; and all the engineers in Santa Rosa and Colorado Springs who made it a success—Mark Woodward, Randy King, Mike Hart, Tim Bagwell, Don Faller, Joe Straznicky, Naily Whang, Mike McTigue, Chris Duff, Rin Park, Rick Martinez, Greg LeCheminant, and Lorenzo Freschet.

Others who have provided me with insight into innovation over my years at HP and Agilent Technologies include Bob Austin, Dave Bass, Russ Johnson, Sigi Gross, Donna Leever, Neal Buren, Megan Chura, Nigel Mott, Tom Lillig, Willard MacDonald, Mike Van Grouw, Hamid Mashouf, Dick Anderson, Jim Olson, and Clyde Coombs. From my career at TeamLogic IT and beyond, they include Don Lowe, Richard Lowe, Dan Shapero, Frank Picarello, Chuck Lennon, Gary Tousseau, Travis Toomey, and John Rosebaugh.

I'd also like to thank the people who helped in the preparation of this book. Guy Foster, a colleague of mine for many years, reviewed many of the chapters and gave me excellent advice that helped clarify my vision. Helen Chen of Keysight Technologies read my case study of the digital communications analyzer and made sure it did not divulge any company-confidential information. Marc Mayer of Keysight's legal department navigated the corridors of company splits to secure the permission I needed for the photo of the DCA product line. Dr. Albert Churella of Kennesaw State University gave me excellent guidance for my chapter on the transition of locomotives from steam to diesel and pointed out the *Burlington Zephyr* as an early example of MVP. My developmental editor Nicholas Machida and my author success manager Neena Laskowski of Elite Authors have been instrumental in helping take the book from well intentioned to well developed. Corinne Moulder and Sophia Moriarty of Smith Publicity have done a wonderful job of making sure it doesn't just sit unread on a back shelf somewhere.

I would also like to thank my son, Greg Hinch, and my daughter, Juliana Hinch, for the insight they have given me—Greg on the challenges of managing start-ups in high tech and Juliana on the challenges of managing a hybrid workforce in the field of medical education. Finally, I want to give recognition to one person who is no longer here to accept it: my wife, Nicki. Over the course of nearly half a century, she gave me unwavering support for my many ventures and was instrumental in assuring the success of our TeamLogic IT franchise. She will be missed.

About the Author

Stephen W. Hinch has over twenty years' experience in senior management positions serving R&D, marketing, manufacturing, and business general management at Hewlett-Packard and Agilent Technologies. He also served as president of TeamLogic IT in Santa Rosa, California. He has personally led innovation projects that have contributed well over $1 billion of new revenue to those companies. He holds three patents in telecommunications technologies and is the author of the groundbreaking professional reference book *Handbook of Surface Mount Technology*. The international electronics industry trade association IPC International presented him with their President's Award for his outstanding contributions to the advancement of the electronics industry. He now serves as executive consultant to companies in the high-tech industry. In his free time, Steve enjoys exploring remote corners of the American Southwest. His 2022

Photo: Will Bucquoy

book on the canyon country of southern Utah, *The Slickrock Desert*, was named winner of an IBPA Benjamin Franklin Award for being judged one of the best nonfiction books of the year. His website is www.stephen-w-hinch.com.

References

Anthony, Scott D., Paul Cobban, Natalie Painchaud, and Andy Parker. *Eat, Sleep, Innovate: How to Make Creativity an Everyday Habit Inside Your Organization*. Harvard Business Review Press, 2020.

Arora, Ashish et. al. "The Changing Structure of American Innovation: Some Cautionary Remarks for Economic Growth." *Innovation Policy and the Economy, Volume 20*. University of Chicago Press, 2020.

Astorino, Steven. "IBM's Made-in-Canada Incubator Goes Global." *IBM Blogs – Canada* (August 21, 2020). https://www.ibm.com/blogs/ibm-canada/2020/08/watson-labs-global/.

Bathlzer, Matt, and Sid Ramtri. "Three Essentials of Successful Corporate Venture Capital." McKinsey & Company podcast (November 2, 2023). https://www.mckinsey.com/capabilities/strategy-and-corporate-finance/our-insights/three-essentials-of-successful-corporate-venture-capital.

Bender, Emily M. et. al. "On the Dangers of Stochastic Parrots: Can Language Models Be Too Big?" *Proceedings of the 2021 ACM Conference on Fairness, Accountability, and Transparency* (2021): 610–623. https://doi.org/10.1145/3442188.3445922.

Bennett, Eve. "Industry Warms to HP's Ideas on SMD Sizes." *Electronics* (October 21, 1985): 19–20.

Chatterjee, Saikat, and Thyagaraju Adinarayan. "Buy, Sell, Repeat! No Room for 'Hold' in Whipsawing Markets." Reuters (August 3, 2020). https://www.reuters.com/article/business/buy-sell-repeat-no-room-for-hold-in-whipsawing-markets-idUSKBN24Z0XY/.

Christensen, Clayton M. *The Innovator's Dilemma: When New Technologies Cause Great Firms to Fail.* Harvard Business Review Publishing, 2000.

Churella, Albert J. *From Steam to Diesel: Managerial Customs and Organizational Capabilities in the Twentieth-Century American Locomotive Industry.* Princeton University Press, 1998.

D'Urso, Joey. "Sorare: 'An Unregulated Timebomb' or a Fantasy Game That Will Revolutionize Football?" *The Athletic* (Nov. 26, 2021). https://www.nytimes.com/athletic/2972039/2021/11/27/sorare-unregulated-timebomb-fantasy-game-revolutionise-football/.

Denning, Steve. "Microsoft CEO Nadella's Brilliant Depiction of the Digital Age." *Forbes* (January 27, 2022). https://www.forbes.com/sites/stevedenning/2022/01/26/microsoft-ceo-nadellas-brilliant-depiction-of-the-digital-age/.

Hale, Rachel. "She Lost Her Scholarship Over an AI Allegation—and It Impacted Her Mental Health." *USA Today* (January 22, 2025).

Hinch, Steve. "SMT Component Standards Needed Now—A Workable Solution Is Offered." *Circuits Manufacturing* (March 1986): 45–56.

Hinch, Stephen W. *Handbook of Surface Mount Technology.* Longman Scientific and Technical, 1988.

Hinch, Stephen W., Michael J. Karin, and Christopher M. Miller. "A New Instrument for Waveform Analysis of Digital Communications Signals." *Hewlett-Packard Journal* 47, no. 6 (December 1996): 6–12.

Hyatt, Michael. *The Vision Driven Leader: 10 Questions to Focus Your Efforts, Energize Your Team, and Scale Your Business.* Baker Books, 2020.

Koen, Peter, Ananya Sheth, Mike DiPaola, and Linda Hill. "Scaling Up Transformational Innovations: Lessons for the C-suite." *Harvard Business Review* (November-December 2024): 79–85.

Kumamoto, V. "The Role of University Research Centers in Promoting Research." *Journal of the Academy of Marketing Science* 45 (2017): 453–458.

Levitt, Theodore. "Exploit the Product Life Cycle." *Harvard Business Review* (November 1965).

Meyer, Marc H., and Alvin P. Lehnerd. *The Power of Product Platforms: Building Value and Cost Leadership.* Free Press, 1997.

Moore, Geoffrey A. *Inside the Tornado: Strategies for Developing, Leveraging, and Surviving Hypergrowth Markets.* Harper Business; Reissue edition, 2005.

Moore, Geoffrey A. *Dealing with Darwin: How Great Companies Innovate at Every Phase of Their Evolution.* Penguin Group, 2005.

Moore, Geoffrey A. *Zone to Win: Organizing to Compete in an Age of Disruption.* Diversion Books, 2015.

Norton, Hugh S. "The Locomotive Industry in the United States, 1920–1960, a Study in Output and Structural Changes." *The Railway and Locomotive Historical Society Bulletin* (October 1965): 66–76.

Packard, David. *The HP Way: How Bill Hewlett and I Built Our Company.* Harper Business, 1995.

Packard, David, and William Hewlett. "Hewlett-Packard Statement of Corporate Objectives." *Measure Magazine* (July 1974).

Porter, Michael. *Competitive Advantage: Creating and Sustaining Superior Performance.* Free Press, 1998.

Rayna, Thierra, and Ludmila Striukova. "The Curse of the First-Mover: When Incremental Innovation Leads to Radical Change." *International Journal of Collaborative Enterprise* 1, no. 1 (2009). http://ssrn.com/abstract=1404015.

Rempel, David et. al. "Reliability of Open Public Electric Vehicle Direct Current Fast Chargers." *SSRN* (April 7, 2022). https://dx.doi.org/10.2139/ssrn.4077554.

Ries, Al, and Jack Trout. *Marketing Warfare 20th Anniversary Edition.* McGraw Hill, 2012.

Rinker, Brian. "A Peek Inside the Hidden, Messy World of Corporate Venture Capital." *Stanford Graduate School of Business Insights* (January 20, 2022).

Schaninger, Bill, Bryan Hancock, and Emily Field. *Power to the Middle: Why Managers Hold the Keys to the Future of Work.* Harvard Business Review Press, 2023.

Strebulaev, Ilya, and Amanda Ying Wang. "Organizational Structure and Decision-Making in Corporate Venture Capital." *Stanford Graduate School of Business Insights* (January 20, 2022).

Tabrizi, Behnam. *Going on Offense: A Leader's Playbook for Perpetual Innovation.* IdeaPress Publishing, 2023.

Tedlow, Richard S. "The Struggle for Dominance in the Automobile Market: The Early Years of Ford and General Motors." *Business and Economic History, Second Series, Volume Seventeen.* The Business History Conference, 1988.

US Copyright Office. "Copyright Registration Guidance: Works Containing Material Generated by Artificial Intelligence." *Federal Register* 88, no. 51 (March 16, 2023).

Vasquez-McCall, Belkis et. al. "How CEOs Are Turning Corporate Venture Building into Outsize Growth." *McKinsey Digital* (October 22, 2024). https://www.mckinsey.com/capabilities/mckinsey-digital/our-insights/how-ceos-are-turning-corporate-venture-building-into-outsize-growth.

Venkatasubramanian, Venkat, and Arinit Chakraborty. "Quo Vadis ChatGPT? From Large Language Models to Large Knowledge Models." *Computers & Chemical Engineering* 192 (January 2025).

Yerramilli-Rao, Bobby et. al. "Strategy in an Era of Abundant Expertise." *Harvard Business Review* (March-April 2025).

Index

Accenture Innovation Centers, 126

Acquisitions, *see* Mergers and Acquisitions

Administrative Assistants, 17, 31, 95, 96

Adobe Reader, 69

Agentic AI, 206

Agile Methodology, 138, 140–143, 147

Agilent Technologies, xiii, 32, 40, 62, 90, 94, 98, 99, 126, 150, 168–170, 181, 186, 187
company split in 2014, 40

Alcatel, 46, 47, 125

Alcatel-Lucent, 125

Alco, 105–108, 111, 112, 114–116, 118–120, 183, 196

Allred, Loren, 150

AlphaCode, 200

Amazon, 84, 89, 126, 207
Amazon Lab, 126, 126
innovation in, 9, 89

Amazon Advantage, 89

Amazon Marketplace, 89

American Locomotive Company, *see* Alco

Anderson, Dick, 25

Android Phone, 6, 145, 196, 197

Apple Inc., 3, 6, 12, 19, 28, 34, 39, 89, 163, 164, 166, 174, 196

Arora, Ashish, 127

AT&T, 46, 47, 125

Automobile Manufacturing, 9, 109, 144

Bain & Company, 128

Baldwin Locomotive Works, 106–108, 111, 114–117, 119–121

Bass, Dave, 98

Bell, Alexander Graham, 125

Bell Labs, 125

Betamax Video Standard, 44, 194, 196

Big Boy Locomotive, 105

Blackberry (smartphone), 6, 12

Blue-Sky Innovation, 90, 95, 101, 126, 133, 135

BMW, 10, 194

Bolivia, 201

Boston Consulting Group, xvi

BP, 6, 11

Budd, Edward G., 113

Budd, Ralph, 113

BudLight Project, 50–61

  achievement of market leadership, 53

  attempt by Colorado Springs to recover, 60

  introduction of, 57

  market and competitive reaction to, 53–55, 57, 61, 62

  name origin, 52, 53

Burlington Zephyr, 113, 114, 140

Business Unit, 26, 28, 36, 126, 127, 129, 131–133, 135, 148, 149

Cadillac, 160

Carbody, 111, 117

Carbon Systems, LLC, 65–78, 97, 182

  creation of, 70

  hiring of remote workers, 74, 75

  innovation strategy, 70–74

  organization design, 74, 75

  relationship with Asus, 76

  relationship with Intel, 71–74, 76

  relationship with MSPs, 67–74

  target markets, 66–70

Centipede Locomotive, 117

Central Research Laboratory, 125

Chatbot, 203, 204

ChatGPT, 97, 200, 203, 205, 207

Chevrolet, 160

Chevron, 6

Chicago, Burlington and Quincy Railroad, 113

Chicago, IL, 113, 140

Christensen, Clayton, 10, 123

Churella, Albert J., 104, 107, 111, 115, 118, 119

Circuits Manufacturing (magazine), 192, 193, 195

Cisco Systems, 47, 136

Civil War (American), 105

Claude AI, 200, 205

Clock Recovery Receiver, 59

CMIT Solutions, 67

Combined Charging System (CCS), 194, 196

Company Objectives, 83

Consultants, x, xii, 65, 162, 177, 181, 191
    what they do well, xvi
Cook, David, 66, 74, 77, 88
Coors Light Project, 60
Copyright Protection, 205, 206
Core Values, 83, 85
Cornell University, 118
Corporate Average Fuel Economy (CAFE), 145
Corporate New Business Creation Group (NBC group), 132, 133, 135, 136, 147, 148
Corporate Venture Capital, 128–132, 147, 148, 186
COVID-19 pandemic, 81
Creeping Featurism, 59
Cryptocurrency, 89
C-Suite, 32, 38, 129, 135, 168, 170–172, 174, 187
Curse of the Corporate Business Model, 23, 25, 28, 36, 37, 45, 100, 120, 208
Customer Advisory Board, 76, 168

DCA, *see* Digital Communications Analyzer
DCA-J, 151, 154, 155, 180, 181, 183, 185
Deepfake (GenAI), 208
DeepSeek, 205, 206
Dell, 39, 65, 67, 68, 71, 76, 78

Denver, CO, 113, 140
Dickerman, William, 119
Diesel Flammability, 111
Diesel Locomotive, 111–118, 120, 124, 140, 177
    A-unit vs. B-unit, 116
    first diesel locomotive, 111
    FT locomotive, 115, 116
    mainline freight engine, 115
    percentage in service, 104
    preference over gasoline power, 111
    road switcher, 118
    switch engine, 114
Digital Age, xiii, 97, 199, 200, 209
Digital Communications Analyzer (DCA), 52–54, 57, 60, 62–64, 90, 94, 97, 99, 124, 125, 131, 145, 146, 149–151, 154, 155, 165, 168, 178–183, 185, 187
    BudLight project, 53
    name origin, 53
Digital Sampling Oscilloscope, 22, 25, 31, 32, 36, 37, 41, 42, 44–46, 48–51, 53–57, 59–61, 179
Discord, 97
Disney, 27
Disruptive Innovation, xii, 8, 10–13, 28, 29, 34, 36, 37, 63, 79, 81, 82, 95, 100, 103, 119, 120, 123–129, 132, 133, 135, 136, 138, 146–150, 155, 156,

159, 161, 162, 164, 176–179,
183, 184, 187, 188, 190, 193,
195–197, 202, 205, 206
definition, 10, 11
organization structure and
management, 125–136
in the railroad industry,
103–121
rules for success, 136–146
Doodlebug, 108
Dropbox, 9, 97
Duke University, 127

E-book Reader, 89
Edison, Thomas, 91
Edward G. Budd Manufacturing
Company, 113
Egbert, Perry T., 118–120
Einstein, Albert, 7
Electric Locomotive, 104, 107,
108, 119
Electric Vehicle (EV), 3, 121,
177, 194, 196
Electro-Motive Corporation
(EMC), 109, 110, 112, 114–
116, 118, 120, 124
Electro-Motive Division of
General Motors (EMD), 112,
115–120, 196
Employee Training, 95, 189
European Parliament, 206
Excavator, 123

F-35 Lightning II, 126
Fiber-Optics, 42–44, 48, 55–58,
124, 140, 150, 165, 181, 196
market segments, 46
market size, 47
SONET/SDH industry
standards, 46, 48, 55–57
Finance, 3, 26, 92, 96, 98
Florida East Coast Railroad, 109
Focus Group, 168–170
Ford, Henry, 144, 159, 160–162
Ford Motor Company, 159
*From Steam to Diesel: Managerial
Customs and Organizational
Capabilities in the Twentieth-
Century American Locomotive
Industry* (book), 104
Frost & Sullivan, 62
FT Locomotive, 115, 116

Gartner, 66
Gasoline Flammability, 111
GenAI, 97, 200–207, 209, 210
agentic AI, 206
hallucination, 204
General Electric (GE), 107–112,
118, 119, 183
General Motors (GM), 112, 113,
119, 120, 159, 160, 194
acquisition of EMC, 112
as world's largest car company,
159, 160

Generative AI, *see* GenAI
Gilligan's Island, 57
GitHub Copilot, 200
Google, 6, 10, 27, 39, 69, 97, 126,
    200, 203, 209
    X Development LLC
    (Google X), 126
Google Chrome, 69
Google Drive, 97
Great Depression, 112, 115
Great Recession of 2007–2008,
    32, 187

Hallucination (GenAI), 204
Hamilton, Harold L., 109, 110,
    112, 113, 117, 120
Hewlett, Bill, 40, 41, 62, 84,
    93, 94
Hewlett Packard Enterprise, 126
Hewlett-Packard Company,
    xi, 13, 21, 39–64, 83, 84, 191
    Boise, 123
    Colorado Springs, 41, 42
    company split in 2000, 40
    HP Labs, 125
    Santa Rosa, 42, 51
    Vancouver, 123
High-Contrast Management, xiii,
    xiv, 37, 146, 208
Hinch, Greg, ix, xv, 130, 190
Hitachi, 194
HP Inc., 21, 65, 126

HP Labs, 43, 125
Hyperloop, 202

IBM Area 631, 136
iBooks, 89
Ideation, 15
iMac, 12
Incremental Innovation, 8, 10–13,
    18, 28, 29, 34, 37, 81, 82, 92,
    100, 101, 103, 119, 125, 133, 146,
    150, 161, 176, 178, 182, 183, 187
    definition, 10, 82
    six methods to encourage,
    81–100
Industry Standardization, 193,
    198
Inkjet Printer, 123
Innovation, ix–xv, 1, 3–19, 21, 25, 26,
    28, 29, 31, 32, 34–37, 39, 63, 65,
    70, 72, 73, 77, 81–83, 85, 88–98,
    100–103, 119–121, 125, 126,
    131, 132, 135–138, 146, 148, 155,
    157, 159, 161, 162, 164, 165, 173,
    175, 176, 178, 183, 184, 186, 187,
    189, 197, 199–202, 208–210
    blue-sky, 133
    as a competitive weapon, 17, 31
    corporate venture capital
    approach, 128–131
    creating environment for
    incremental innovation, 82
    definition, 4

determining innovation type,
29, 30
disruptive, *see* Disruptive
Innovation
environment for innovating, 31
ideation vs. innovation, 31
importance of collaboration, 5
incremental, *see* Incremental
Innovation
influencing the industry,
189–197
market driven, 26
by market phase, 175–188
process, 18
product, 18
radical, 10
in remote organizations, 77
senior management support, 137
service, 4
seven steps to success, 34, 35
sustaining, 10
tenets, 18
types, 18
value of separate innovation
department, 82
world's most innovative
companies, 6
The Innovator's Dilemma (book),
10, 123
Internet, 10, 43, 149, 150, 199, 200,
207, 209, 210
iPad, 28, 86–89, 99

iPhone, 3, 6, 12, 39, 145, 196
iPod, 12, 164, 166

Jasper AI, 200
JCPenney, 23
Jitter, 154
Jitterfest, 154, 155
Jobs, Steve, 34, 161, 163
Johnson, Kelly, 126, 136
Johnson, Russell, 57

Kaufman Act of 1923, 111, 121
Kettering, Charles F., 113, 120
Keysight Technologies, 40, 52,
62, 126
Kindle, 89
Kodak, 13
Kohl's Innovation Center, 126

Lab Stock, 94
Land Cruiser, 145, 146
Large Language Model (LLM),
200, 204, 205, 207, 209
Laser Printer, 123, 189
Laser Transmitter, 43, 44, 58
Lehnerd, Alvin, 143
Lenovo, 65, 67, 68, 76, 78
Life Cycle (Market), 14, 173, 175,
184, 187, 198, 200
Light Saber Project, 58
Lima Locomotive Works, 106, 107,
116, 117

Lisa Computer, 12, 163
Lockheed, 23, 126
Lockheed Martin, *see* Lockheed
Lowe's Innovation Labs, 126

Macintosh Computer, 12
Main Street Market, 150
Managed Services Provider (MSP),
ix, 67–71, 73, 74, 77, 78, 85
challenges working with computer
manufacturers, 67–69
role in computer warranty
support, 69
services provided, 67
as trusted IT advisor, 67
Management, xi–xvi, 3, 4, 7, 11, 16,
17, 21–25, 28, 31–38, 42, 45,
49–53, 57–63, 74–76, 81, 82,
84, 85, 87, 88, 90–96, 98–100,
102, 109, 110, 114, 118–120,
125, 129, 136, 137, 139, 142,
149, 151, 155, 156, 166, 174,
175, 177, 178, 181, 182, 185,
187, 208
C-suite, 32
in family-run business, 34
importance of empowered
management, 32
managers as customers, 168
mid-level, 32, 155
role in innovation, 5
senior-level, 32, 129

tradeoffs in decision making, 22
Manager, *see* Management
Manufacturing, 8, 10, 14, 17, 26,
27, 39, 82, 84, 90, 92, 97, 98,
106, 110, 116, 123, 136, 145,
146, 185, 189, 191, 192
automobile, 109, 144, 145
manufacturing engineers, 17
Market Phases, 175–188
decline 183, 184
emerging, 177
growth, 177–179
mature, 179–183
Market Research, 49, 54, 62, 161,
163, 165, 168, 173, 174
customer advisory board, 168
focus group, 169
online methods, 168
phases of market, 175, 176
segmentation, 165–167
Market Segmentation, 4, 47, 165,
167, 173
business-to-business markets, 166
business-to-consumer markets,
165
template, 47, 166, 167
Marketing, 14, 16, 17, 22, 26, 27,
32–34, 42, 44, 53, 54, 57, 70, 76,
82, 84, 85, 92, 97–99, 109, 110,
118, 119, 136, 150, 151, 154, 155,
165, 168, 181, 183, 185, 200
customer analysis, 46

customer research vs. market research, 165

market research, *see* Market Research

oscilloscope market segments, 41

in railroad industry, 100

Marketing Warfare (book), 70, 76

Matsushita, 194

McKinsey & Company, 128

Mergers and Acquisitions, 27, 28, 107, 129–131, 184–187

Meyer, Marc, 143

Microsoft, 5, 11, 12, 39, 68, 69, 97, 166, 199, 200

Microsoft Research, 126

Microsoft Copilot, 97, 200

Microsoft Office, 69

Microsoft Teams, 97

Minimum Viable Product (MVP), 138–140, 147

Mitsubishi, 194

Model A, 160

Model T, 159, 160

Moore, Geoffrey, 7, 136, 149, 177, 181, 188, 191

Motorola, 6

MP3 Player, 163, 174

MSP, *see* Managed Services Provider

MVP, *see* Minimum Viable Product

Nadella, Satya, 199

National Association of Broadcasters Trade Show, 99

National Bureau of Economic Research, 127

National Institute of Standards and Technology (NIST), 154

New York City, 111, 121

"The Next Boulder," 153, 154

Nobel Prize, 125

Nokia, 6, 125

Nortel, 46, 47

North American Charging Standard (NACS), 194–196

NUC, 71, 72, 74, 76

Odin project, 54, 139

185th Aero Squadron, 118

Oscilloscope, 40–46, 48, 49, 52–57, 60–62, 151

invention of, 40

market segments, 41

market size, 41

price, 42

see also Digital Sampling Oscilloscope

P-38 Lightning, 126

Packard, Dave, 40, 41, 62, 84, 93

Palm, 12

Patents, 10, 15, 16, 19
  importance of, 15
  process, 10
Patton, George, xv
Personal Computer, 12, 13, 19, 39,
  42, 45, 123, 199, 207, 209
Photodetector, 43–46, 48, 50, 53,
  56, 57, 60, 125
Photoshop, 13
Porsche Consulting Innovation
  Lab, 126
Procter & Gamble, 9, 166
Product Platform, 143
Prompt (GenAI), 200, 203, 204, 209

R&D, ix, 3–5, 7, 14–18, 21, 22, 26,
  27, 33, 34, 42, 49, 53, 57–59, 74,
  82, 92, 94, 97, 98, 126, 135, 136,
  144, 150, 151, 154, 169, 185, 186
  innovation in, 14–16
  research vs. development,
  7, 127
Railcar, 108–113
Research and Development, 3
Reuters, 23
RIM, see Blackberry
Rosebaugh, John, 74, 77, 97

Salesforce, 27
Sampling Scope, see Digital
  Sampling Oscilloscope
Schaffer, Charlie, 42, 53, 181
Sears, 23

Senior Leaders, xiii, xiv, 4, 32, 33,
  83, 85, 138, 155, 171, 172, 177,
  181, 195, 208
  role in innovation, 34, 35, 83
Sharp Corporation, 194
Shell Oil, 6, 11
The Silver Streak (1934 movie), 113
Skunk Works, 126, 136
Slack, 97
Sloan, Alfred P., 113, 160
Small and Midsize Business (SMB)
  Market, 36, 65, 66
Smith Corona, 23
SMTA (Surface Mount Technology
  Association), 192, 195
SONET/SDH standards, 46–48,
  50, 51, 55–58, 60, 64
Sony, 9, 12, 163, 174, 194, 196
Sorare, 6
Southern Pacific Railroad, 109
SpaceX, 89, 202
SR-71 Blackbird, 126
Stage-Gate Inc., 142
Stanford Graduate School of
  Business, 130
Star Trek, 202
Start-ups, xiv, 11, 24, 25, 27, 36, 54,
  59, 65, 127–132, 135, 148, 175,
  177, 184–188, 213
  relationship with corporations,
  127, 128, 130, 131
Steam Locomotive, 104–107, 110–
  112, 115, 117, 119–121, 124, 177

growth in horsepower, 105
mainline steam, 105
percentage in service, 104
switch engine, 111
transition to diesel, 115, 117
Steam Shovels, 123
Sullivan, Bill, 32, 33, 150
Swift, Taylor, 96

Target, xiii, 23, 47, 50, 66
TeamLogic IT, ix, x, xiii, 66, 67,
    70, 74, 85, 88, 213
Tektronix, 40–42, 44–46, 48–50,
    53–62, 64, 178, 179, 181
    company formation, 40
    oscilloscope for fiber-optics
    market, 42, 48
Tesla, 3, 34, 194–198
Test and Measurement Business,
    21, 22, 40
Through-Hole Technology, 189, 190
Toomey, Travis, 88
Tornado Market, 149, 150, 177,
    181, 207
Toyota, 26, 145, 160
Toyota 4Runner, 145, 146
Toyota New Global Architecture
    (TNGA), 145, 146
Trainset, 113, 114, 140
Transformer-Based GenAI, 202
Treo, 12
Tuck, John, 192
Turner, Lincoln, 45, 181

The 20 MSP, 67
2001: A Space Odyssey (movie), 201

Union Pacific Railroad, 105
United Airlines, 9
US Copyright Office, 206

Venture Capital, 11, 27, 65,
    128–130, 132, 133
VHS video standard, 44, 194
Video Production Companies, 99
Volkswagen Automotive
    Innovation Lab, 126

Walkman, 163
Wall Street, 23
Walmart, 23
Waterfall Methodology, 142
Westinghouse, 107, 108, 109
Whitcomb Locomotive Works,
    115, 121
White Automobile Company, 109
White House (Washington, DC), 66
White Mountain (California), 152
Wikipedia, 200, 201
Winton Engine Company,
    110, 112
World War I, 116, 118
World War II, 116, 118
World's Fair of 1939, 110

Zone to Win (book), 136
Zoom, 49, 64, 75, 97

www.ingramcontent.com/pod-product-compliance
Lightning Source LLC
Chambersburg PA
CBHW022113210326
41597CB00047B/295